ADVANCES IN URETHANE SCIENCE AND TECHNOLOGY

VOLUME I

Edited by:

K. C. Frisch and
S. L. Reegen

Polymer Institute
University of Detroit

a TECHNOMIC® publication
TECHNOMIC Publishing Co., Inc.
750 Summer St., Stamford, Conn. 06901

ADVANCES IN URETHANE SCIENCE AND TECHNOLOGY

VOLUME I

750 Summer St., Stamford, Conn. 06901

. TECHNOMIC® publication

PRINTED IN U.S.A.

Library of Congress Catalog Card No. 75-150348

Standard Book No. 87762-062-8

ADVANCES IN URETHANE SCIENCE AND TECHNOLOGY

CONTENTS

FOREWORD

This volume is the formal proceedings of the 1970 Polymer Conference Series held at The Polymer Institute of The University of Detroit.

The purpose of the Polymer Institute's Conferences is to present the latest developments in the field of urethane research, including the newest production processes. Accordingly, this is also the purpose of this volume.

The papers cover catalysis, property-structure relationships, elastomers, flame-resistant foams, integral skin foams, sandwich panels, urethane floors, and other aspects of urethane technology.

It is the hope of the editors of this volume that further advances in urethane technology and chemistry will be documented by subsequent volumes of this type.

The Polymer Institute
The University of Detroit
Detroit, Michigan
April 6, 1971

Kurt C. Frisch
Sidney L. Reegen

CHAPTER 1

CATALYSIS IN ISOCYANATE REACTIONS

INTRODUCTION

Catalysis plays an important role in the reactions of isocyanates. It is of vital importance in the manufacture of urethane and urethane-urea polymers in the form of foams, elastomers, coatings, sealants, adhesives, etc. Catalysis not only affects the rates of the chemical reactions responsible for the formation of these polymers but it also exerts a considerable influence on their ultimate properties.

The catalysis of isocyanate reactions can generally be grouped into two categories:

1. Polymerization of isocyanates.
2. Reaction of isocyanates with active hydrogen-containing compounds.

POLYMERIZATION OF ISOCYANATES

The formation of cyclic dimers and trimers from aryl isocyanates was reported by Hoffman [1] well over 100 years ago. However, the formation of linear polymers has been reported only as recently as 1959 [2, 3].

The general mechanism proposed for polymerization of isocyanates consists of formation of an active complex by attachment of a nucleophilic catalyst, X, onto the carbon of the isocyanate group with displacement of an electron pair to the nitrogen atom, followed by addition of another isocyanate molecule:

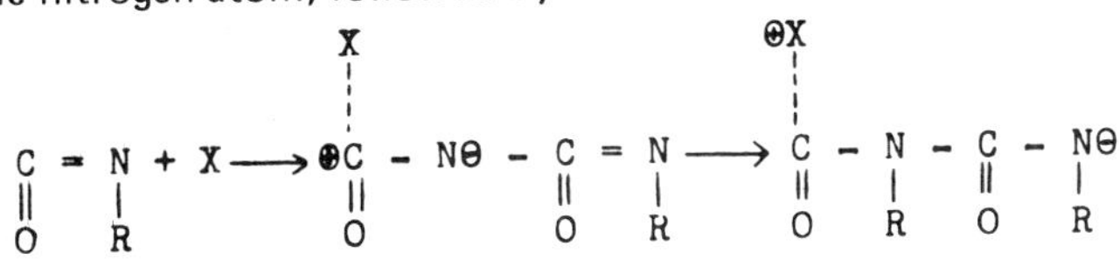

The relative rates of subsequent reactions determine whether a dimer, trimer or a linear polymer is formed. If chain termination occurs at this stage with loss of X, a cyclic dimer is formed. If a third monomer molecule is added, followed by loss of X, a cyclic trimer is obtained. If chain termination is relatively slow, addition of further monomers takes place with formation of a linear polymer.

DIMERIZATION

Aromatic isocyanates have a tendency to dimerize:

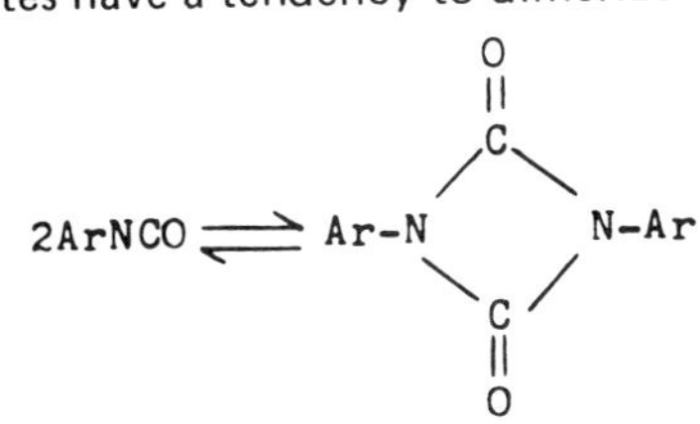

The rate of self polymerization of isocyanates to dimers (uretidine diones) depends upon the electronic or steric influences of ring substituents. Ortho substitution greatly retards dimerization of the NCO group. Hence, the NCO group in the 2-position of tolylene diisocyanate (TDI) is slower to dimerize than the NCO group in the 4-position. Dimerization is catalyzed strongly by trialkyl phosphines [4, 5] and more mildly by tertiary amines such as pyridine [6, 7]. 4,4′-Diphenylmethane diisocyanate (MDI) dimerizes slowly on standing at room temperature even without catalysts but is stable at low or at slightly elevated temperatures (40-50°C).

Isocyanate dimerization is an equilibrium reaction. Dissociation of the dimer occurs only at elevated temperatures [8]. In the absence of catalysts, temperatures as high as 175°C. are required to completely dissociate the dimer of 2, 4-TDI although initial dissociation was observed to occur at 150°C [9]. Aliphatic isocyanate dimers have so far not been obtained and only trimers have been isolated.

TRIMERIZATION

In contrast to dimer formation, trimerization is not an equilibrium reaction since trimers are stable in the range of 150° to 200°C:

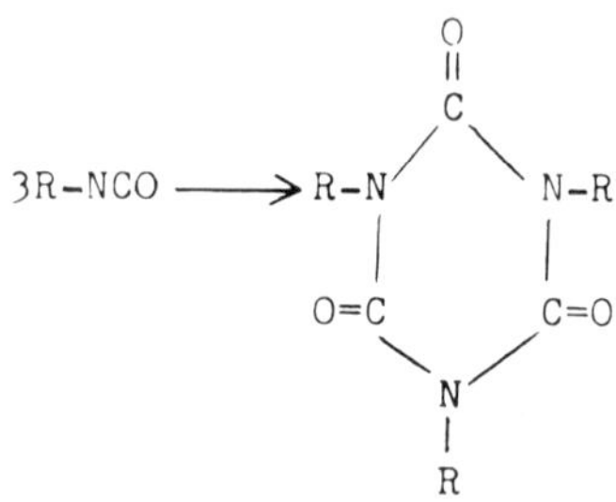

Catalysts for trimerization of aliphatic and aromatic isocyanates include calcium acetate [5], potassium acetate, sodium carbonate [10], sodium methoxide [11], triethylamine, oxalic acid, sodium benzoate in dimethyl formamide [12], soluble compounds of iron, potassium, magnesium, mercury, nickel, copper, zinc, chromium, aluminum, tin, vanadium, titanium [13], triethyl phosphine [14], titanium tetrabutyrate [15] and oxygen [16].

Kogon [17] found that the reaction between phenyl isocyanate and ethyl alcohol at 125°C. in the presence of N-methylmorpholine led to the formation of the trimer in excellent yields.

In addition, epoxides, particularly in the presence of tertiary amines [18] are effective catalysts. It has been postulated [18] that the active species is an amine-epoxide complex having a zwitterion structure:

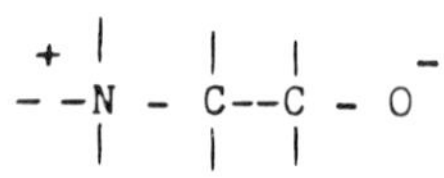

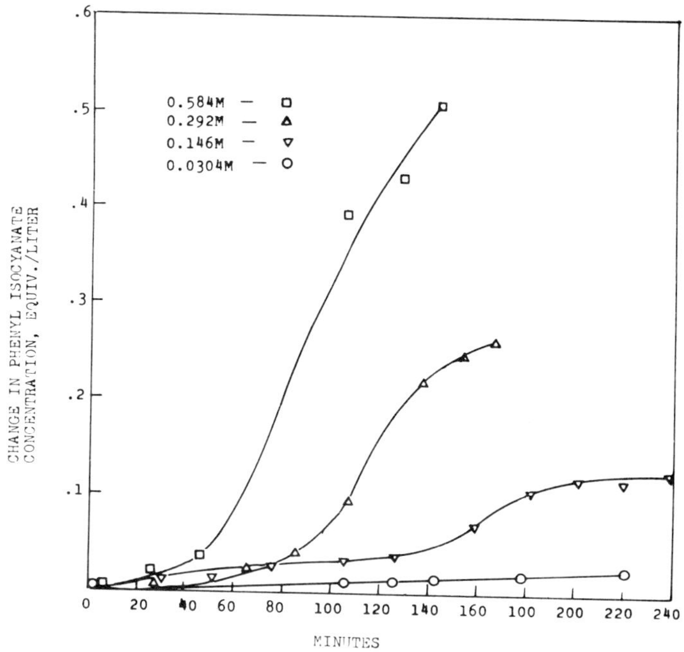

Figure 1. Polymerization of Phenyl Isocyanate with (0.0032M) Triethylenediamine and (0.0286M) Propylene Oxide in Benzene Solution at 26-27°C.

Beitchman [19] found that a triethylenediamine-propylene oxide combination was significantly more active than any other amine-propylene oxide combination. The polymerization of phenyl isocyanate, using this catalyst pair, is shown in Figure 1 [19]. In addition to the zwitterion structure of this catalyst combination other possible explanations for the catalytic nature of this reaction are:

1. Formation of an unstable complex of olefin oxide and isocyanate due to the catalytic effect of triethylenediamine.
2. Formation of an unstable molecular complex of triethylenediamine, olefin oxide and isocyanate.

In addition to epoxides, other cocatalysts found to be effective were lower molecular weight alkylenimines or aldehydes [19].

The preparation of isocyanurate-containing polyurethanes was reported by Burkus [20] who employed an isocyanate-terminated polyester prepolymer and a combination of triethylamine and propylene oxide as trimerizing agent.

The preparation of isocyanurate-containing polyurethane foams with improved high temperature properties was described by Nicholas and Gmitter

[20-A]. These investigators also studied the relative effectiveness of trimerization of phenyl isocyanate using a number of catalysts listed in Table 1. The effect of solvents on the rate of trimerization of phenyl isocyanate is given in Table 2. These data indicate that solvents having high dielectric constants increase the rate of the trimerization reaction.

The effect of substituents on the trimerization of phenyl isocyanate using N,N',N''-tris (dimethylaminopropyl) sym-hexahydrotriazine (HHT) as catalyst in chlorobenzene solution is shown in Table 3 (20-A). These results indicate that electron attracting groups substituted on the phenyl ring accelerate trimerization while electron donating groups retard the reaction. Steric factors also strongly influence trimerization, e. g., p-chlorophenyl isocyanate is approximately 1400 times as reactive as o-chlorophenyl isocyanate. In o-nitrophenyl isocyanate, the activating inductive and electromeric forces

Table 1. Activity of Catalysts in Trimerization of Phenyl Isocyanate.

	Relative Reactivity
2,4,6 Tris(dimethylaminomethyl) phenol (a)	1.0
2,4,6, Tris(dimethylaminomethyl) phenol: diglycidyl ether of bisphenol-A (1:1)	1.2
Mixture of ortho and paradimethylaminomethyl phenol (b)	1.5
N,N',N'' Tris(dimethylaminopropyl) sym. hexahydrotriazine	2.4
N,N',N'' Tris(dimethylaminopropyl) sym-hexahydrotriazine; diglycidyl ether of bisphenol-A (1.1)	7.7
Benzyl trimethylammonium hydroxide in dimethyl sulfoxide (as 25% solution)	8.5
Benzyl trimethylammonium methoxide	4.0
Sodium methoxide in dimethyl formamide (as saturated solution)	25.0

(a) DMP-30 (Rohm & Haas).
(b) DMP-10 (Rohm & Haas).

Table 2. Effect of Solvents on the Rate of Trimerization of Phenyl Isocyanate*.

Solvent	Dielectric Constant At 20°C.	Relative Reactivity
Benzene	2.28	1.47
Chlorobenzene	5.67	2.4
Acetonitrile	37.5	8.4
Dimethyl sulfoxide	45.0	12.3

*11.9 g.(0.1 M) phenyl isocyanate.
13.1 g. of solvent as above.
0.1 g. of N,N',N'' tris(dimethylaminopropyl) sym-hexahydrotriazine.

(-I, -E) overcome the steric effect since the rate of trimerization is twice that of phenyl isocyanate.

LINEAR POLYMERIZATION

Monoisocyanates can polymerize to linear, high molecular weight polymers which are termed N- substituted 1-nylons. The structure of these polymers can be represented as:

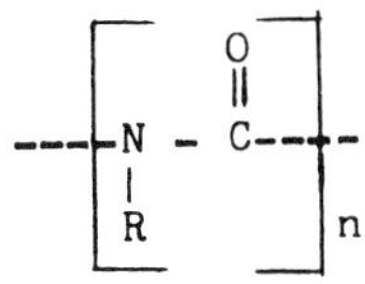

R may be alkyl or aryl groups and n may be up to 2000.

The polymerization is generally carried out by treatment of solutions of monoisocyanates in solvents such as N, N-dimethylformamide with a soluble anionic catalyst such as sodium cyanide, sodium metal, sodium naphthalene or sodium benzophenone ketyl at low temperatures (-20°to -100°C).

In the initiation step, the anion (X-) attacks the isocyanate group as follows:

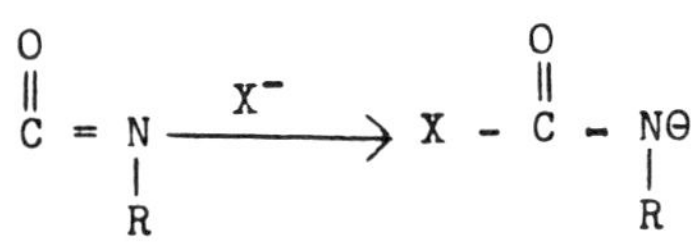

Table 3. Rates of Trimerization of Substituted Phenyl Isocyanates

Substitutent	Relative Reacivity
Unsubstituted	10.0
o-Methyl	0.06
m-Methyl	1.92
p-Methyl	3.95
o-Chloro	0.09
m-Chloro	70.0
p-Chloro	128.0
p-Bromo	144.0
o-Nitro-	20.5
m-Nitro	fast
p-Nitro-	fast
o-Ethoxy-	slow—not measurable
80/20 mixture of 2,4 and 2,6 tolylene diisocyanate	5.4

The nitrogen anion will propagate the reaction until a linear trimer stage is reached:

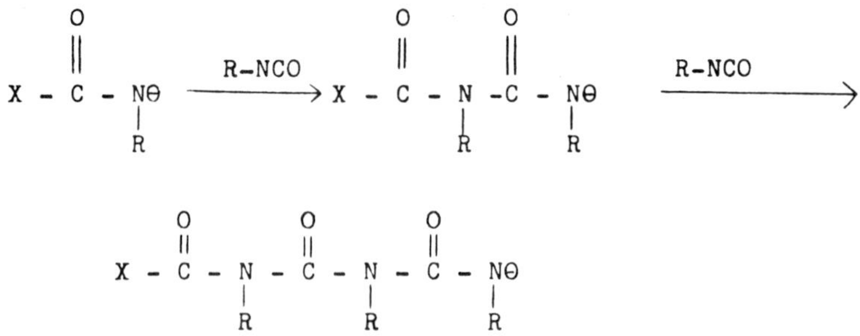

The linear trimer can follow two reaction paths depending upon the reaction conditions. At high concentrations of initiator and at high temperatures, formation of a cyclic trimer occurs. If the initiator concentration is low, the monomer concentration is high and the polymerization is carried out at low temperatures, a linear polymer will result:

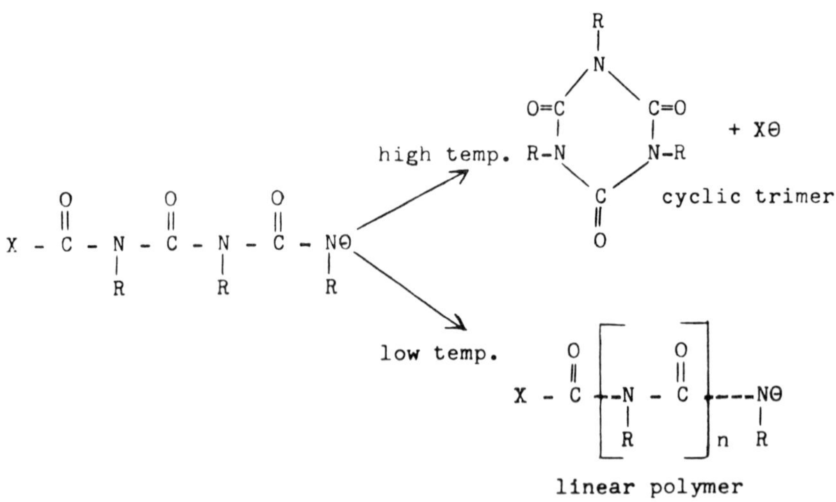

The termination step presumably occurs during the isolation of the polymer and (or) by the reaction of traces of water present in the reaction mixture:

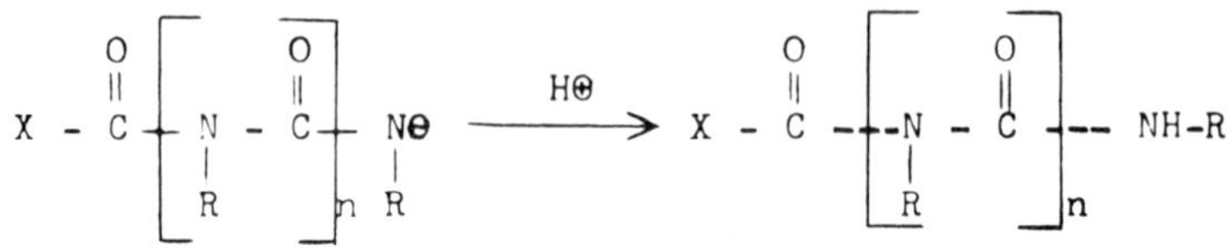

CARBODIIMIDES

Carbodiimides, containing the characteristic linkage –N=C=N–, are obtained by means of a number of non-catalytic reactions, e.g., heating isocyanates at elevated temperatures [21-22] or by thermal degradation of polyurethanes [23].

It is interesting to note that one member of the phosphine family, 3-methyl-l-phenyl-3-phospholine oxide, can bring about quantitative conversion of diisocyanates to polycarbodiimides [24-27]:

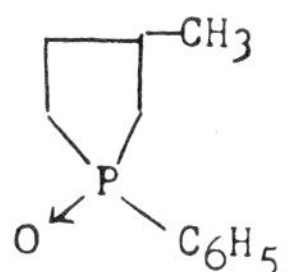

$$2R-NCO \longrightarrow R-N=C=N-R + CO_2$$

Likewise phospholines and phospholidines have been reported to catalyze carbodiimide formation [28].

The catalytic action of a selective phosphine, 3-methyl-1-phenyl-3-phospholine, on phenyl isocyanate brings about diphenylcarbodiimide formation in nearly 90% yield [8]. This is the more remarkable since most of the phosphine catalysts are very effective catalysts for dimer formation and carbodiimide formation can usually be considered merely a side reaction when isocyanates are treated with phosphine catalysts.

It has been found that the resistance to hydrolysis of polyester-based urethane elastomers may be significantly improved by the addition of small amounts of polydicarbodiimides [29].

REACTIONS OF ISOCYANATES WITH ACTIVE HYDROGEN-CONTAINING COMPOUNDS

The catalysis in the reaction of isocyanates with active hydrogen compounds is often greatly complicated by the fact that one or more side reactions may occur and that the presence as well as choice of catalyst may have a profound

Polyurethanes, although characterized by the typical linkage $-NH-\overset{O}{\overset{\|}{C}}-O-$(urethane), may contain many other functional groups within their polymer molecule. These may include ether, ester, urea, amide, allophanate, biuret, and other groups.

The basic reactions in the formation of various types of polyurethanes, are briefly reviewed below. For simplicity, the reactions in these equations are presented in most instances in their monofunctional form.

The two most important reactions in the preparation of polyurethanes are the reactions between isocyanate and hydroxyl-containing compounds (polyether or polyester polyols) and the reaction between isocyanate and water. The former is the basic reaction for the formation of urethane groups and can be considered as the chain propagating reaction:

$$R-N=C=O + R'-OH \rightarrow R-\underbrace{NH-\overset{O}{\overset{\|}{C}}-O}_{\text{urethane}}-R'$$

The second reaction is responsible for the foaming of the urethane polymer by the liberation of carbon dioxide with simultaneous formation of substituted urea groups. The first step in this reaction is the formation of the unstable carbamic acid which decomposes to form an amine and carbon dioxides:

$$R-N=C=O + H_2O \rightarrow \underbrace{\left[R-NH-\overset{O}{\overset{\|}{C}}-OH\right]}_{\text{carbamic acid}} \rightarrow R-NH_2 + CO_2$$

The amine immediately reacts with additional isocyanate to form a substituted urea as follows:

$$R-NH_2 + R-N=C=O \xrightarrow{\text{fast}} R-\underbrace{NH-\overset{O}{\overset{\|}{C}}-NH}_{\text{urea}}-R$$

The overall reaction can be represented as:

$$2R-N=C=O + H_2O \longrightarrow R-NH-\overset{O}{\overset{\|}{C}}-NH-R + CO_2$$

An alternate mechanism provides for formation of a carbamic acid anhydride, followed by decomposition to a substituted urea and carbon dioxide:

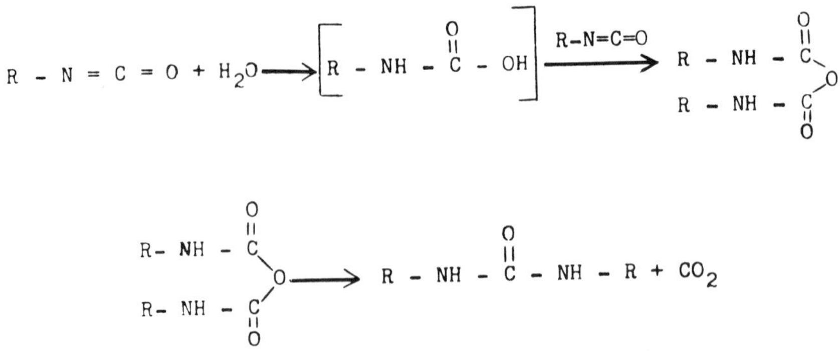

Still another reaction which may occur to some extent is the formation of the amine salt of carbamic acid which decomposes under the influence of heat to yield water and urea:

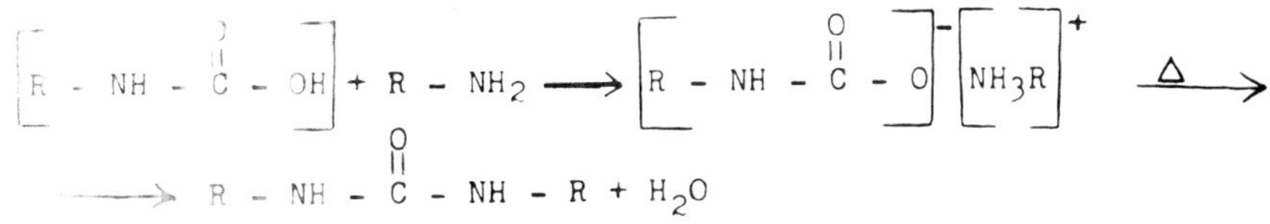

Other reactions which lead to branching and crosslinking in urethane foams are the formation of allophanate and biuret linkages:

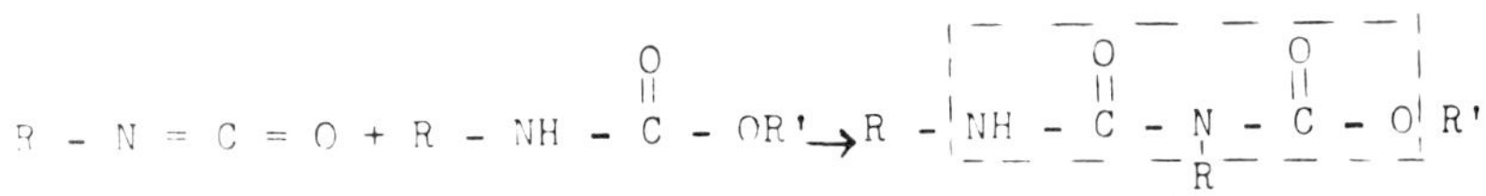

allophanate

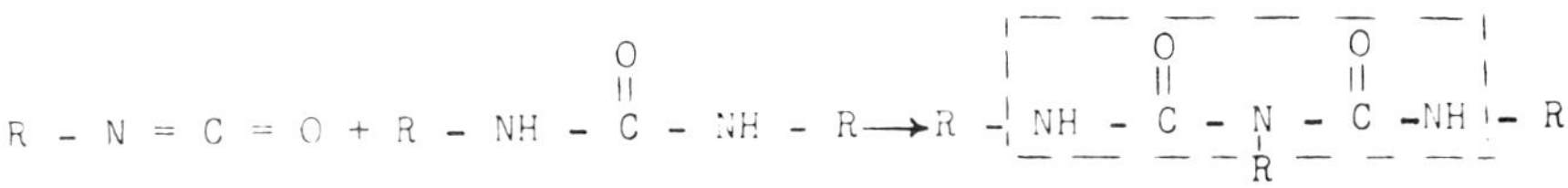

biuret

Still another crosslinking reaction is trimerization, leading to the formation of isocyanate rings which are hydrolytically and thermally very stable:

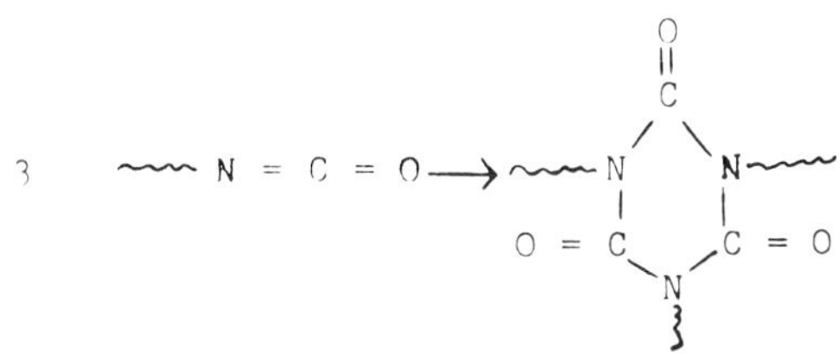

Since most of the above reactions are too slow for the commercial manufacture of urethane foams, catalysts are employed whose functions are not only to bring about faster rates of reaction but also to establish a proper balance between the chain propagation reaction (mainly the hydroxyl-isocyanate reaction) and the foaming reaction (isocyanate-water reaction in the case of flexible foams). A balance has to be established between gas formation and polymer growth in order to entrap the gas (CO_2) efficiently and to develop sufficient strength in the cell walls at the end of the foaming reaction (evolution of gas) to maintain their structure without collapse or shrinkage.

Another important function of catalysts in foam reactions as well as in the formation of other types of urethane polymers is to bring about completion of the reactions, resulting in an adequate "cure" of the respective polymer. Completion of cure results in maximum strength properties, and maximum chemical and weathering resistance. Progress of cure can be determined by means of differential thermal analysis (DTA), infrared absorption and Vicat softening point data [30, 31].

The catalysts most commonly employed are tertiary amines and metal catalysts, especially tin catalysts. Tertiary amines are catalysts for both the isocyanate-hydroxyl as well as the isocyanate-water reaction. The efficiency of tertiary amine catalysts depends upon their chemical structure. It generally increases as the basicity of the amine increases and the steric shielding of the amino nitrogen decreases [32, 33]. Table 4 [34] shows the importance of steric factors in the catalytic activity of tertiary amines. Some of the most commonly used tertiary amine catalysts are triethylene diamine ("Dabco", Houdry Process & Chem. Co.), N-alkyl morpholines, N,N,N'N'-tetramethyl-ethylene diamine, N,N,N',N'-tetramethyl-1,3-butanediamine, N,N'-substituted piperazines, and dialkylakanolamines.

Organotin compounds are extremely effective catalysts for the isocyanate-hydroxyl reaction. They include stannous octoate, stannous oleate, dibutyltin dilaurate and dibutyltin di-2-ethylhexoate.

It is common practice to use combinations of tertiary amines and tin catalysts in the production of flexible foams and to some extent in rigid foams requiring a "hot" catalyst system such as for spray foams or for the manufacture of structural panels. In other cases, rigid foams are made employing tertiary

Table 4. Relative Catalytic Activity of Amine Catalysts.

Chemical Name of Amine Catalyst	Chemical Formula	pK_a	Relative Catalytic Activity
1,4-Diazabicyclo(2,2,2)-octane	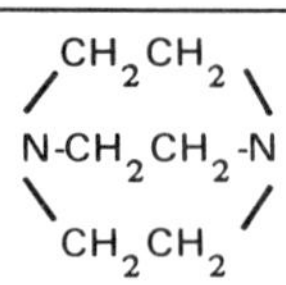	5.4	23.9
N,N-Diethylcyclohexylamine	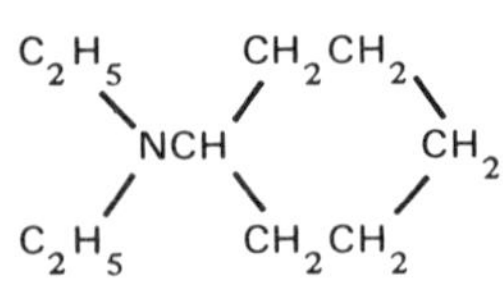	10.0	0.7
N-Methyl-morpholine	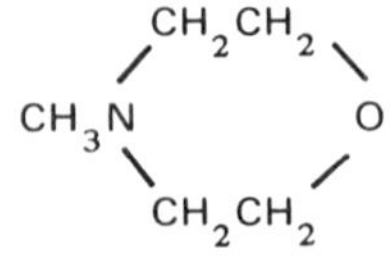	7.4	1.0
N,N,N',N'-Tetramethyl-methane diamine	$(CH_3)_2NCH_2N(CH_3)_2$	10.6	0.085

amine catalysts either singly or in combination (e.g., N,N,N'N'-tetramethylethylenediamine and dimethylethanolamine).

Table 5 shows the relative order of activity of various catalysts and catalyst combinations in a model isocyanate-hydroxyl reaction [35]. It also points out clearly the synergistic effect between tin catalysts and tertiary amines. It is of interest to note that a synergistic effect has also been observed with two tertiary amine catalysts, e.g., tetramethylbutanediamine and triethylamine [36]. This is probably due to a combination of two different factors—relative lack of steric hinderance and effect of relatively strong basicity.

The relative order of activity of some catalysts for the water-isocyanate reaction is seen in Table 6 (35).

Table 5. Relative Reactivity of Catalysts in Isocyanate-Hydroxyl Reaction.

Catalyst	Concentration %	Order of Activity
Uncatalyzed	-	1
TMBDA	0.1	56
DABCO	0.1	130
TMBDA	0.5	160
DBTDL	0.1	210
DABCO	0.2	260
DABCO	0.3	330
SnOct	0.1	540
DBTDL	0.5	670
DBTDL + TMBDA	0.1 + 0.2	700
SnOct + TMBDA	0.1 + 0.2	1000
DBTDL + DABCO	0.1 + 0.2	1000
SnOct + TMBDA	0.1 + 0.5	1410
SnOct + DABCO	0.1 + 0.5	1510
SnOct	0.3	3500
SnOct + DABCO	0.3 + 0.3	4250

TMBDA = Tetramethylbutanediamine
DABCO = 1,4-Diaza(2,2,2)bicyclooctane
DBTDL = Dibutyltin dilaurate
SnOct = Stannous octoate

Table 6. Relative Reactivity of Catalysts in Isocyanate-Water Reaction 0.1 pts. Catalyst per 100 pts. Resin

Catalyst	Order of Activity
Stannous octoate	1.0
N-Ethyl morpholine	1.1
Dibutyltin dilaurate	1.3
Triethylamine	1.5
N,N,N',N'-Tetramethyl-1,3-butane diamine	1.6
1,4-Diaza (2,2,2)bicyclooctane (DABCO)	2.7

MECHANISM AND KINETICS OF CATALYZED REACTIONS OF ISOCYANATES WITH HYDROXYL COMPOUNDS

Baker and Holdsworth [37] were the first to subject amine-catalyzed reactions between isocyanates and alcohols to kinetic analysis in order to elucidate the mechanism of these reactions. The reaction between phenyl isocyanate and methanol in the presence of tertiary amines was found to follow a second order kinetics, first order with respect to the isocyanate, and first order with respect to the alcohol. The second order rate constant k_2 can be expressed by the following relation:

$$k_2 = k_o + k_c [\text{amine}]$$

where k_o is the rate constant for the uncatalyzed reaction and k_c is the catalytic constant characteristic of the activity of the catalyst. They assumed the following reaction steps:

$$\text{ArNCO} + \text{NR}_3 \underset{k_2}{\overset{k_1}{\rightleftharpoons}} \underset{\displaystyle \text{NR}_3}{\text{ArN}\overset{}{\text{CO}}}$$

$$\underset{\displaystyle \text{NR}_3}{\text{ArNCO}} + \text{R'OH} \xrightarrow{k_3} \text{ArNHCOOR'} + \text{R}_3\text{N}$$

Using a stationary state condition the concentration of the complex is given by the expression

$$[\text{complex}] = \frac{k_1 \ \text{ArNCO} \ \ \text{NR}_3}{k_2 + k_3 [\text{R'OH}]}$$

and hence the rate of the catalyzed reaction can be expressed by

$$-\frac{d[\text{ArNCO}]}{dt} = \frac{k_1 k_3 [\text{ArNCO}][\text{NR}_3][\text{R'OH}]}{k_2 + k_3 [\text{R'OH}]}$$

The overall rate constant is then:

$$k = k_o + \frac{k_1 k_3 [\text{NR}_3]}{k_2 + k_3 [\text{R'OH}]} = k_o + k_c [\text{NR}_3]$$

The catalytic properties of a polymeric tertiary amine, poly-4-vinylpyridine, and its copolymers with styrene and methyl methacrylate were compared with that of pyridine in reactions of mono and diisocyanates with alcohols and glycols [38]. It was found that the reaction rate was proportional to the con-

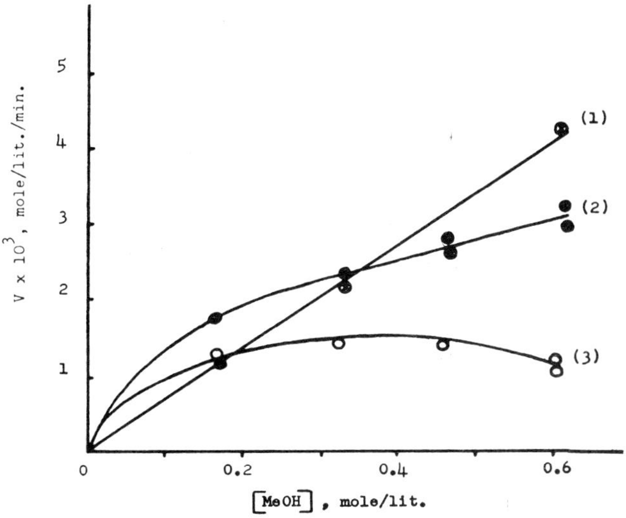

Figure 2. Rate of the Reaction Between PhI and MeOH, Catalyzed (1) by Pyridine, (2) by Copolymer of 4-Vinylpyridine (35.5%) and Styrene (64.5%, and (3) by Poly-4-Vinylpyridine, as a Function of MeOH Concentration in Solution. Concentrations (Moles Per Liter): PhI = MeOH = 0.3; Cat. = 1.5×10^{-2}; T = 40°C.

centration at which the reaction system becomes very viscous which indicates the accessibility of catalytic sites in the polymer for the reactants.

The polymer-catalyzed reactions exhibited higher rates in comparison with those catalyzed by pyridine at low reactant concentrations. However, when the concentration is increased, the rates of the polymer-catalyzed reactions reach limit values and ultimately become less than in the case of pyridine. It was also found that the catalytic activity of the polymeric catalyst increased as the styrene content in the copolymer increased. Figure 2 shows the rate of reaction between phenyl isocyanate and methyl alcohol as a function of the methyl alcohol concentration in solution. Figures 3 and 4 indicate the rates of reaction between TDI and ethylene glycol as a function of the ethylene glycol and TDI concentrations respectively.

The increase in the rate of reactions catalyzed by a polymer in comparison with monomeric catalyst may be due either to higher catalytic activity of active centers or to the increase of the effective concentration of the reactants in the reaction zone. The coincidence of total activation energies of the phenyl isocyanate-methyl alcohol concentration (6 kcal/mole) in the presence of pyridine and of a copolymer (35.5% polyvinyl pyridine/64.5% styrene) as measured in the range of 20° to 60° C. suggests the latter explanation to be the more likely one. These phenomena can be explained by the solution behavior of the polymers which form coils or aggregates that are easily permeated by the reacting molecules. These coils or aggregates have reacting molecules. These coils or aggregates have an ability to dissolve the

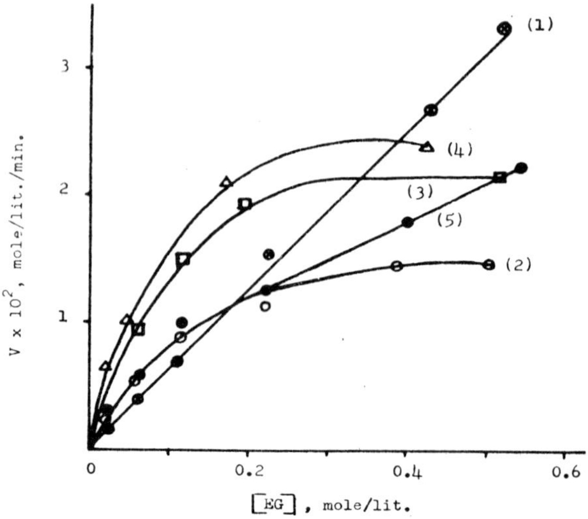

Figure 3. Rate of the Reaction Between TDI and EG as a Function of EG Concentration. Catalysts: (1) Pyridine, (2) Copolymer of 4-Vinylpyridine (50%) and Styrene (50%), (3) Copolymer of 4-Vinylpyridine (35.5%) and Styrene (64.5%), (4) Copolymer of 4-Vinylpyridine (13.5%) and Styrene (86.5%), and (5) Copolymer of 4-Vinylpyridine (35.0%) and Methyl Methacrylate (65%). Concentrations (Moles Per Liter): TDI = 0.2; Cat. = 5×10^{-2}; T = 40°C.

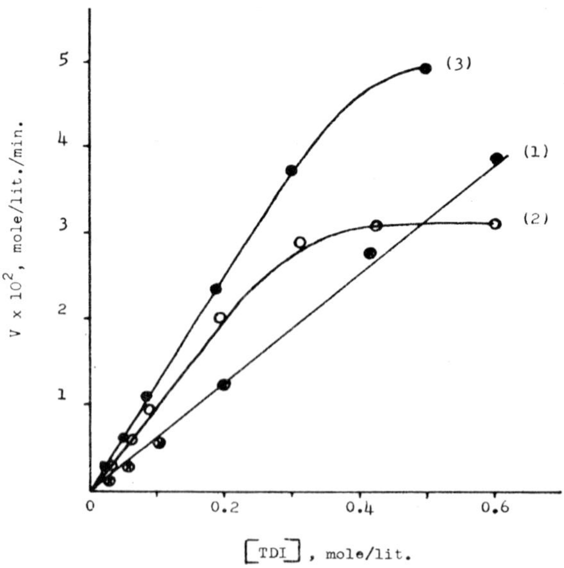

Figure 4. Rate of Reaction Between TDI and EG as a Function of TDI Concentration. Catalysts: (1) Pyridine, (2) Copolymer of 4-Vinylpyridine (35.5%) and Styrene (64.5%), (3) Copolymer of 4-Vinylpyridine (13.5%) and Styrene (86.5). Concentrations (Moles Per Liter): EG = 0.2; Cat. = 5×10^{-2}; T = 40°C.

reactants. It may be assumed that the distribution of the aromatic isocyanate in the solvent (chloroform) and in the polymeric catalyst, containing benzene rings, is shifted in favor of the latter as a result of hydrophobic interaction of the hydrocarbon aromatic residues. Therefore, the isocyanate equilibrium concentration in this type of catalyst would be higher than its average concentration in solution and would increase with the styrene content in the copolymer.

The limiting value of the catalytic activities of the polymeric catalyst can be explained on the basis of saturation of the macromolecular coils or aggregates with reactants.

Hostettler and Cox [39] studied the effect of catalyst concentration on the phenyl isocyanate -1-butanol reaction and found the data to correspond to the following relationship:

$$k_2 = k_o + k_c \,[\text{catalyst}]^n$$

where n=0.89 for di-n-butyltin diacetate and 1.04 for triethylamine. Britain and Gemeinhardt [40] suggested that the metal catalyzed isocyanate-hydroxyl reaction involves a complex formation between the metal compound, the isocyanate and the hydroxyl compound as shown below. The complex formation could occur in two steps with either the isocyanate or the hydroxyl compound reacting first.

In the double complex, the isocyanate and hydroxyl compound are positioned on the same side of the metal compound in close proximity to each other.

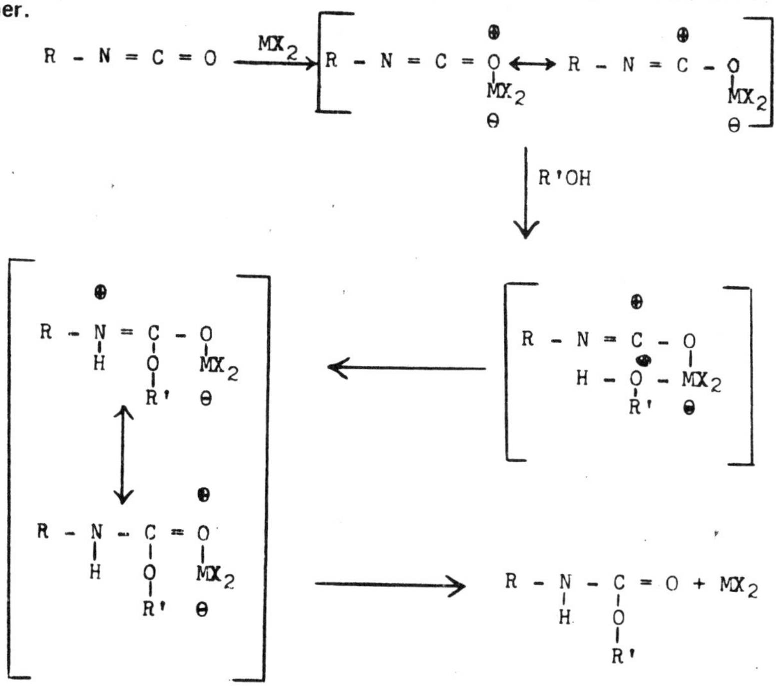

A mechanism for the urethane formation in the reaction between phenyl isocyanate and methyl alcohol in the presence of metal acetylacetonates has been proposed by Lipatova and Nizelskii [41]. They followed the course of reaction employing copper acetylacetonate, an active catalyst for urethane formation [42], in dioxane solution using spectrophotometry and titration of samples for isocyanate. In order to determine whether interaction between the reactants and catalyst occurred, EPR spectra of copper acetylacetonate were taken at 40°C in the presence of either methyl alcohol or of phenyl isocyanate. It was found that the EPR spectrum of copper acetylacetonate was altered only in the presence of phenyl isocyanate. A change in color of the copper acetylacetonate solution in dioxane from blue to green was also observed about 10-15 minutes after the addition of phenyl isocyanate. The same color change took also place in the reaction mixture when methyl alcohol was present. The observed decrease in transmission and the bathochromic shift of the transmission maximum may be explained by the formation in the system of products of the reaction between copper acetylacetonate and phenyl isocyanate. The color of transition metal ions and their complexes is usually connected with d-electron transitions. The change of color to green may be explained by the reaction of the catalyst with phenyl isocyanate which brings about changes in the d $\rightarrow$ d transition of the copper ion Cu^{++}. This only occurs if Cu^{++} is bonded to the phenyl isocyanate. Likewise, a study of the UV spectra of copper acetylacetonate and its reaction product with phenyl isocyanate resulted in bathochromic effects and significant hyperchromic effects in the maximum at 242 mμ, characteristic of the ligand-metal transition [43]. Additional evidence for the immediate interaction between phenyl isocyanate and the copper ion Cu^{++} could also be obtained by the observation of the mixture of peaks of the Π-Π^* transitions of the ligand rigns at 292mμ. In view of the tendency of the copper ion Cu^{++} to form complexes of octahedral symmetry as well as planar complexes, it can be assumed that the observed color change is related to the formation of an octahedral complex having the following structure:

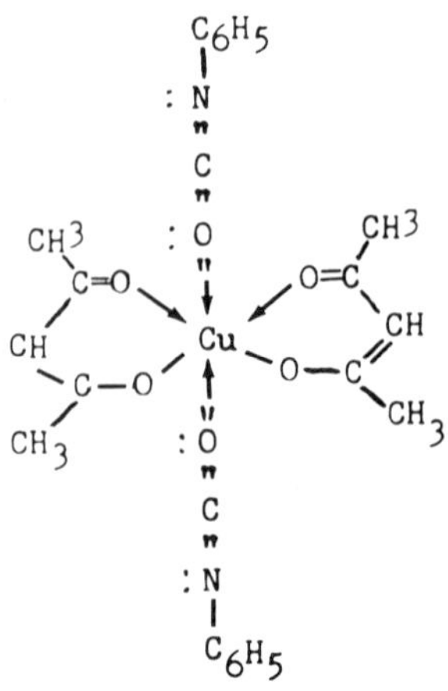

In this complex, the electrophilic nature of the carbonyl carbon of the phenyl isocyanate is increased which enhances the nucleophilic attack of the oxygen atom of the methyl alcohol to form the phenyl urethane. The latter is apparently coordinated with the copper ion Cu^{++} at first and is then separated as the reaction product after decomposition of the complex (II):

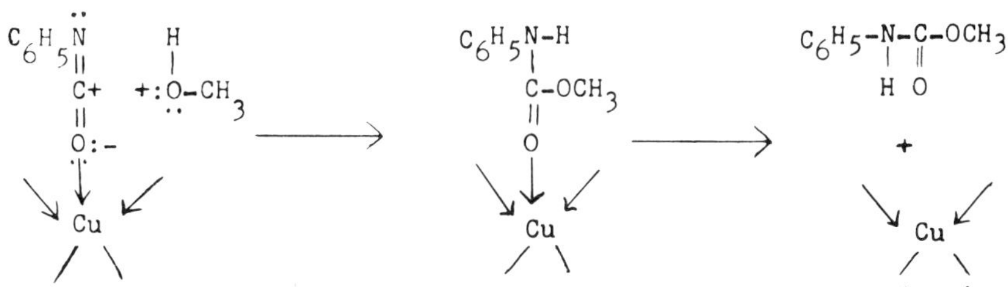

(I) (II) (III)

The fact that the transmission further decreases with time has been ascribed to formation of more complicated complexes. One such product, a dianilide of copper acetylacetonate (IV), was actually isolated when the reaction was carried out in chloroform.

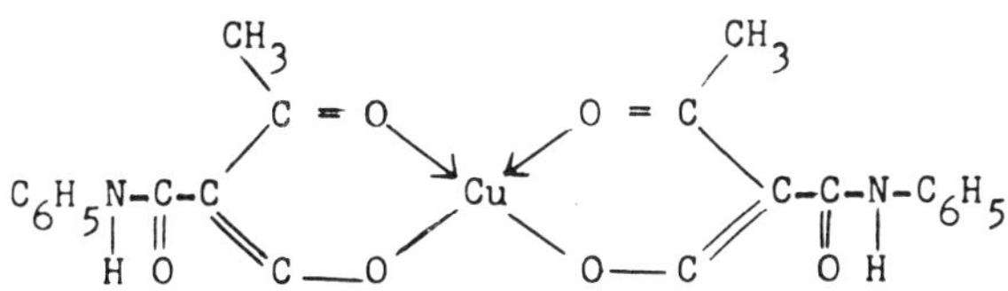

(IV)

Reegen and Frisch [44] noted an apparently constant value for the reaction rate in the reaction between 1-methoxy -2- propanol and phenyl isocyanate in toluene solution at 30°C catalyzed by dibutyltin dilaurate, until about 40-50% disappearance of isocyanate groups was achieved. At that point a noticeable increase in the rate constant was observed. (see Figure 5). The increased rate of isocyanate disappearance could have been caused by the catalytic effect of urethane groups on further reaction of alcohol and isocyanate as had been suggested by Baker and Holdsworth [37]. It could also be due to side reactions involving the isocyanate groups such as formation of trimers or reaction with urethane groups to form allophanates:

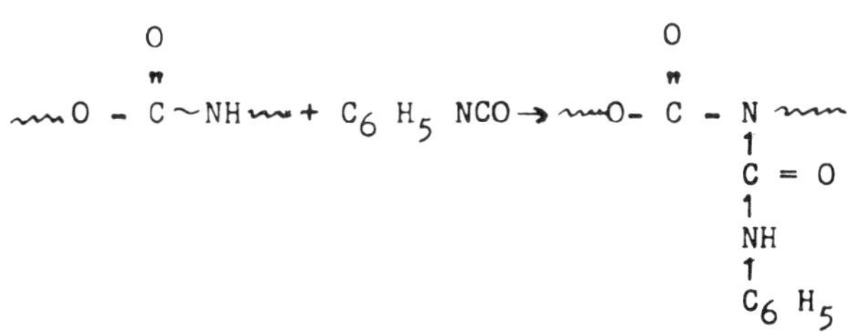

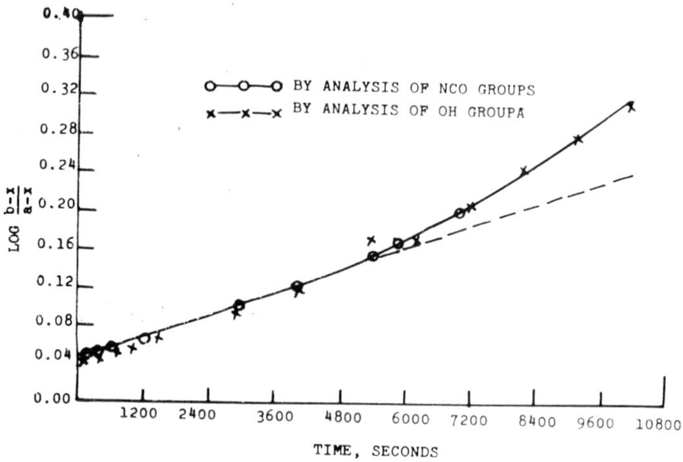

Figure 5. Reaction of Phenyl Isocyanate and 1-Methoxy-2-Propanol, $1X_{10}^{-4}$ Mole/1, Dibutyltin Dilaurate, Toluene, 30°C.

The catalytic effect of urethane groups in the reaction between isocyanates and alcohols (as measured by disappearance of isocyanate) has previously been observed by Okada and Iwakura [45] and Sato [46]. The former have shown that the uncatalyzed reaction of phenyl isocyanate with urethanes does not proceed at a measurable rate at 40°C.

On the other hand Kogon [47] has demonstrated the formation of allophanates by reaction of phenyl isocyanate with ethyl carbamate at room temperature in the presence of various metal catalysts.

In order to determine the mechanism responsible for the apparent increase in the reaction rate, the progress of the reaction was followed by measurement of the disappearance of the hydroxyl groups, employing the method of Gutnikov and Schenk [48]. If a similar increase in rate were noted after 40-50% reaction, it would lend support to previous evidence of the catalytic influence of urethane groups; if no increased rate were observed, it would indicate that side reactions involving isocyanates were playing a major role. The data obtained from both analytical techniques as shown in Figure 5, clearly indicate that increases in rates occur at 40-50% reaction. Hence this effect appears to be due to the increasing concentration of urethane groups formed during the alcohol-isocyanate reaction and the catalytic effect of these urethane groups on the further alcohol-isocyanate reaction.

Smith [49] reported similar observations on a broader variety of reactants and catalysts including mono-and polyfunctional reactants and tertiary amine, cobalt and tin catalysts. The data for the reaction between 1-ethyl-1-2-hydroxybutoxy)-2-butoxy ethane (M-11) and phenyl isocyanate in toluene at 35.56°C., as seen in Figure 6, verified the catalytic effect of the urethane groups formed during the reaction. However, the point in the reaction where this catalysis becomes noticeable as indicated by deviation of the kinetic

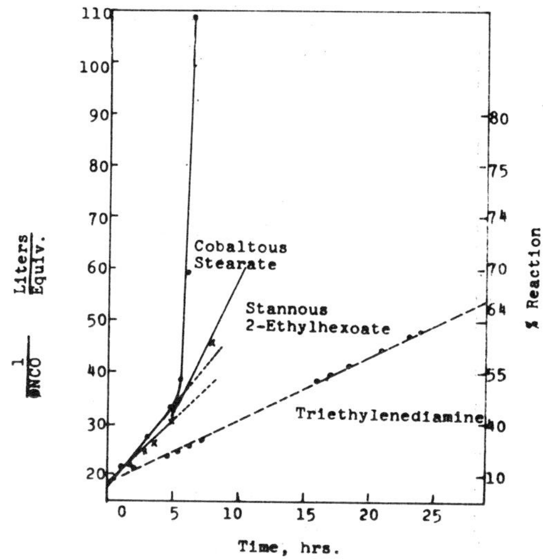

Figure 6. Reaction of Alcohol M-II with Phenyl Isocyanate with Various Catalysts (49).

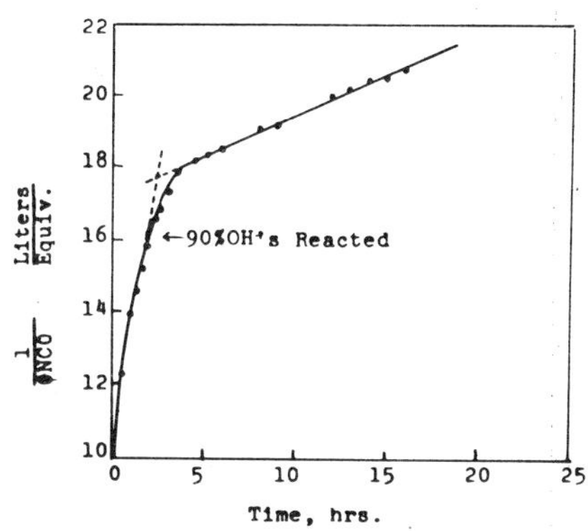

Figure 6a. Reaction of 1-Butanol with Excess Phenyl Isocyanate (49).

plot from linearity, appears to vary with the catalyst used. In the case of stannous 2-ethyl-hexoate, deviation occurs at 37-40% reaction, with cobaltous stearate at 45-50% reaction while with triethylenediamine the deviation occurs only after more than 65% reaction has taken place.

Smith [49] also observed that the type of hydroxyl group, primary or secondary, or the presence of other functional groups in the molecule, such as ethers, do not appear to influence the point where the deviation starts. The sole effect of these groups is merely reflected in the respective rate. (see Figure 6).

An interesting effect was noted by Smith [49] in the use of excess isocyanate in the reaction with monofunctional alcohols. Reegen and Frisch [44],

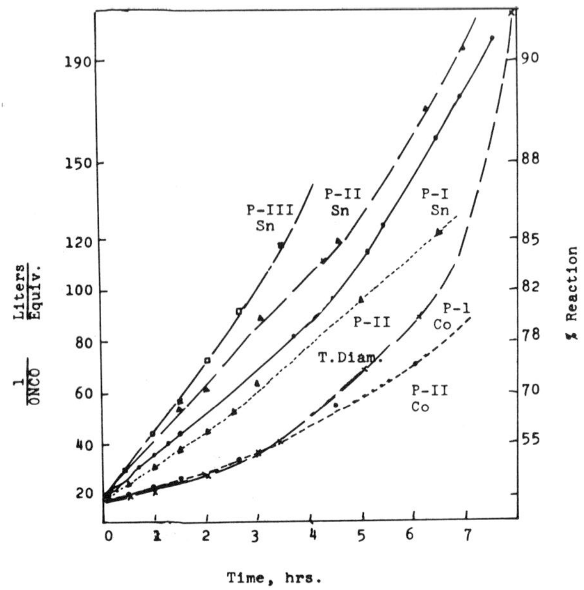

Figure 7. Reaction of Polyhydroxy Alcohols with Phenyl Isocyanate (49).

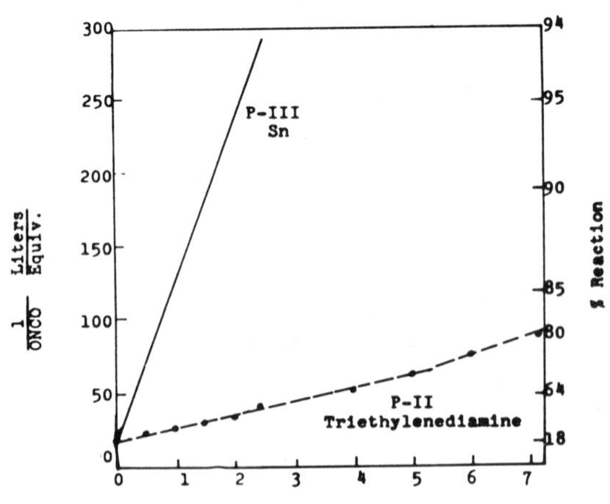

Figure 8. Reaction of Polyhydroxy Alcohols with Tolylene Diisocyanate (49).

employing a 10% excess of isocyanate obtained essentially the same results as Smith using equivalent amounts of reactants. However, if a 100% excess of phenyl isocyanate is used with 1-butanol, using triethylene diamine catalyst, a linear plot is obtained up to 90% reaction of the hydroxyls. Afterwards, the rate decreases to that characteristic for the reaction of the isocyanate group with a urethane group to form an allophanate linkage. In this case no rate acceleration due to urethane catalysis of the alcohol-isocyanate reaction was observed. (See Figure 6-A)

When polyether polyols containing several hydroxyl groups (P-I=polyether polyol with only sec. hydroxyl groups; P-II=polyether polyol with 50% prim. and 50% sec. hydroxyl groups; P-III=amine-based polyol with 50% prim. and 50% sec. hydroxyl groups) were used instead of monofunctional alcohols, data, as shown in Figure 7 were obtained [49]. The tertiary amine-catalyzed reaction is similar to that of the monohydroxyl alcohol reaction, showing deviation from linearity at 60-70% reaction. In the case of cobaltous stearate, deviation occurs at 34-40% with the polyhydroxyl alcohols versus 45-50% reaction with the monofunctional alcohol. Stannous 2-ethylhexoate shows deviation at 80-85% reaction versus 34-40% for the reaction with the monohydroxyl alcohol.

Substituting a diisocyanate such as tolylene diisocyanate for phenyl isocyanate, the data shown in Figure 8 are obtained [49]. The tertiary amine-catalyzed reaction is similar to that of the reaction with the mono-isocyanate, indicating deviation from linearity at 72-75% reaction. However, the tin-catalyzed reaction showed no deviation from linearity up to >93% reaction. reaction.

Smith [49] explained the behavior of the tertiary amine-catalyzed reactions which did not appear to be affected by polyfunctionality of the alcohols in the light of the mechanisms for amine-catalyzed isocyanate reactions. Since it has been demonstrated repeatedly [37, 50, 51] that in these reactions activation of the isocyanate group occurs and that one activated isocyanate group reacts with one hydroxyl, polyfunctionality should not affect the results. However, in the case of a cobalt catalyst, more than one hydroxyl, ether, urethane, or isocyanate group can be bound in a tetrahedral complex (see Figure 9).

The behavior of the tin-catalyzed reactions can also be explained on the basis of the geometry of the reactive complexes formed as shown in Figures 9A and 9B [49].

In the course of a study of the kinetics of alcohol and polyol-isocyanate reactions using metal catalysts, the interesting observation was made that the relative rates of the reactions of model hydroxyl compounds (methoxypropanols) with phenyl isocyanate were reversed when a lead catalyst was used instead of a tin catalyst [52-54]. The reaction rates observed with dibutyltin dilaurate catalyst in toluene solution at 30°C. are shown in Figure 10. As could be expected, both the position of the methoxy group and the nature of the hydroxyl group had a pronounced effect on the reaction rates. Using

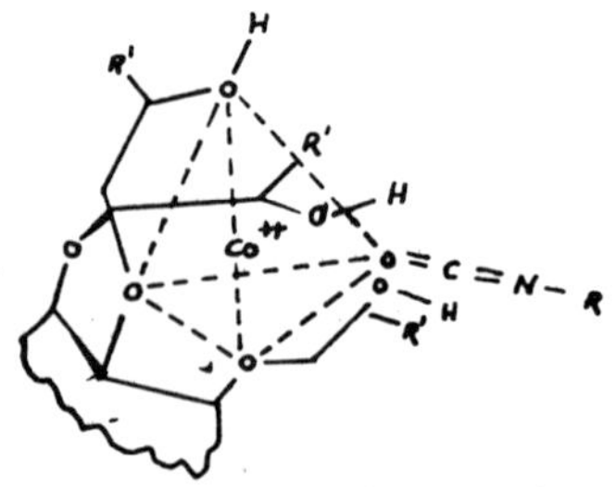

Figure 9. Cobaltous Complex (49).

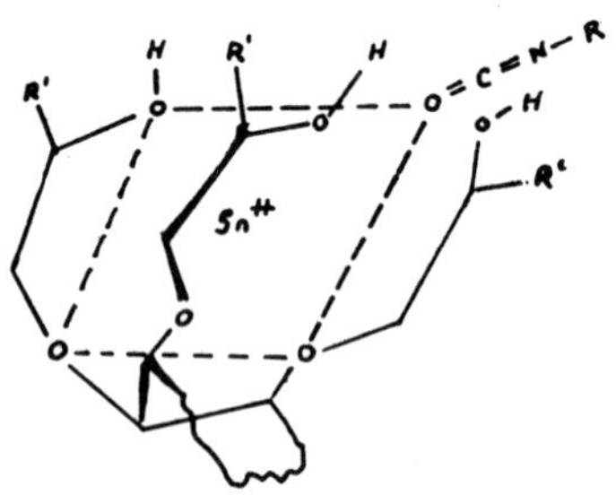

Figure 9a. Stannous Complex (49).

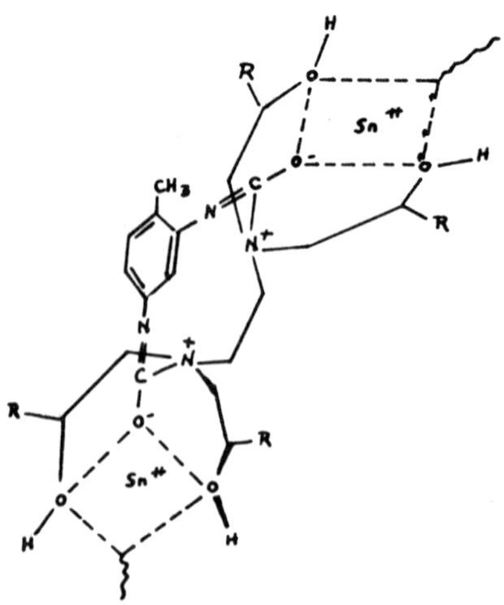

Figure 9b. Polyol P-III and Tolylene Diisocyanate Under Tin Catalysis (49).

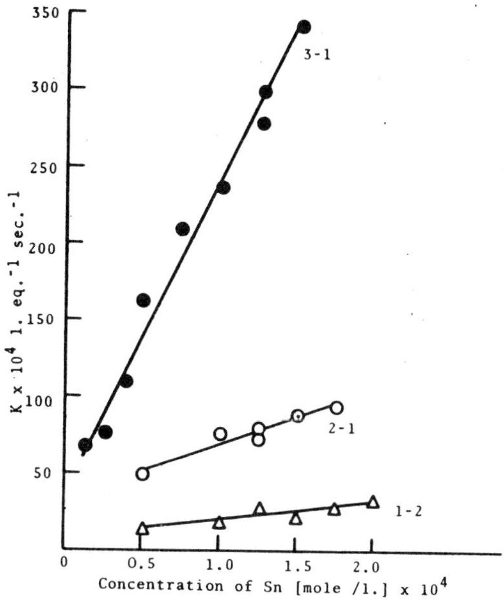

Figure 10. Effect of Sn conc. on Second Order Constant. Reaction of Phenyl Isocyanate with 1-Methoxy, 2-Propanol (1-2), 2-Methoxy, 1-Propanol (2-1), and 3-Methoxy, 1-Propanol (3-1). In Toluene at 30°C.

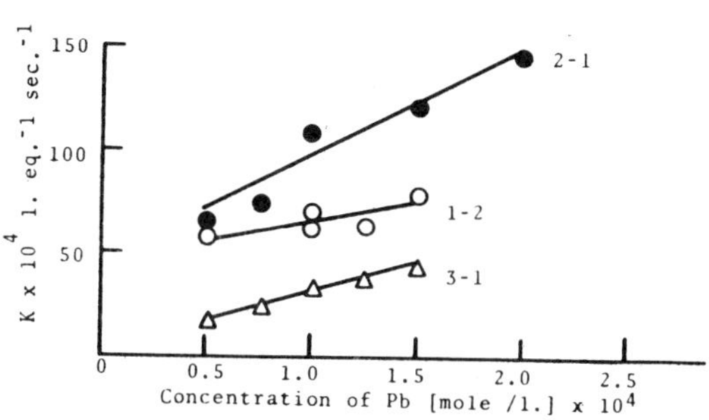

Figure 11. Effect of Pb Concentration on Second Order Rate Constant. Reaction of Phenyl Isocyanate with 1-Methoxy-2-Propanol (1-2), 2-Methoxy-1-Propanol (2-1), 3-Methoxy-1-Propanol (3-1). In Toluene at 30°C.

dibutyltin dilaurate, 3-methoxy-1-propanol exhibited the fastest rate, followed by 2-methylyl-1-propanol and 1-methoxy-2-propanol. However, when employing lead naphthenate as catalyst, 2-methoxy-1-propanol had the fastest rate, followed by 1-methoxy-2-propanol with 3-methoxy-1-propanol exhibiting the slowest rate (see Figure 11). The same trend in reaction rates was observed when either cellosolve acetate or toluene was used as solvent although the rates were considerably slower in the case of cellosolve acetate. This is as would be expected in the case of a polar solvent involving presumably interaction (complex formation) between the polar solvent and the alcohol.

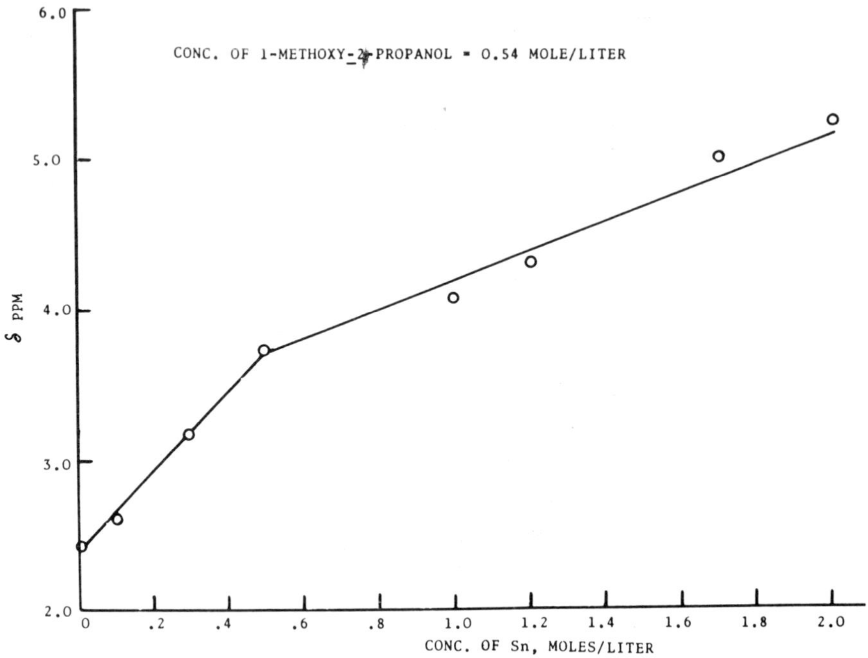

Figure 12. Effect of Sn Concentration on Shift of -OH by NMR.

A study of the ultraviolet spectra in the range of 220-375 mμ mixture of phenyl isocyanate with tertiary amines (tri-n-propylamine and triethylamine) and with metal catalysts (stannous octoate and lead naphthenate) failed to disclose any evidence of isocyanate-catalyst complexes [54].

However recently, Entelis and coworkers [61] during a study of the mechanism of the effect of organotin catalysts on the reaction of p-chlorophenylisocyanate and methanol, reported UV data (using $\lambda = 218$ mμ, which gave evidence for the existence of alcohol-tin catalyst complexes.

On the other hand, shifts in the OH proton resonance in the NMR spectra indicated clearly complex formation for the system dibutyltin dilaurate: 1-methoxy-2-propanol as shown in Figure 12 [54]. It can be seen that the peak associated with the –OH group shifted downfield with increasing tin concentration which may be readily interpreted on the basis of complex formation between the catalyst and the alcohol, resulting in electron withdrawal from the hydroxyl group. The shifts measured were downfield from tetramethylsilane which was used as an internal standard. It is evident from Figure 12 that a 1:1 complex is formed; a sharp decrease in slope of the plot of chemical shift vs. catalyst concentration is obtained beyond the addition of equimolar concentrations of Sn to the alcohol. The continued increase in chemical shift beyond that concentration may be due to the existence of additional complexes.

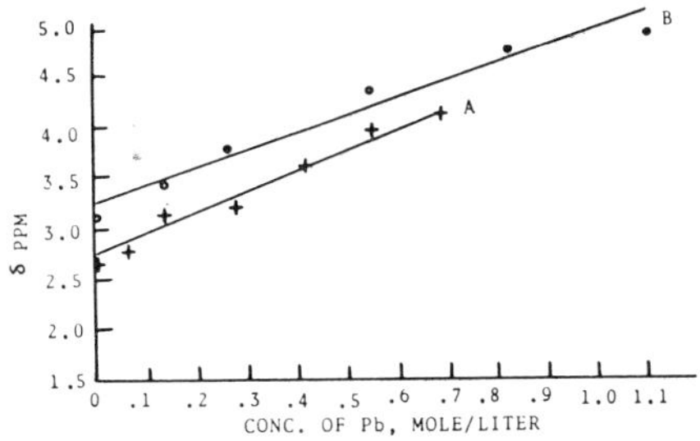

Figure 13. Concentration Dependence of the OH NMR Resonance of 1-Methoxy-2-Propanol on Lead Naphthanate. Concentration: (A) TMS Used as Internal Standard, CCl_4 Solution, Concentration of ROH = 0.53 Mole/1. (B) Cyclohexane Used as Solvent and Internal Standard Corrected to TMS by the Addition of 86 cps. Concentration of ROH = 0.56 Mole/1.

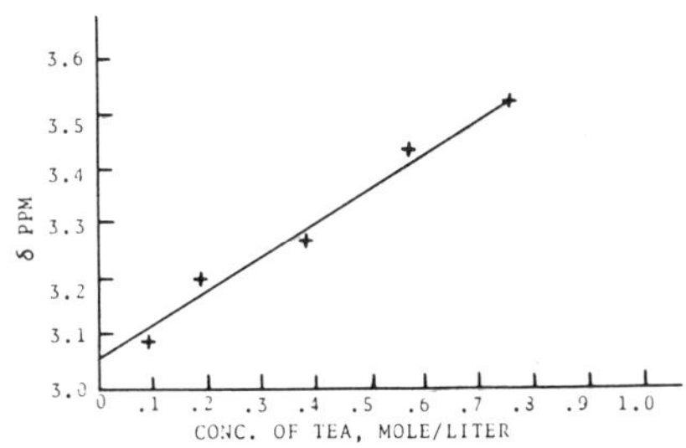

Figure 14. Concentration dependence of the OH proton of 1-Methoxy-2-Propanol on Triethylamine Concentration. Cyclohexane was used as Solvent and Internal Standard, Corrected to TMS by the Addition of 86 cps. Concentration of ROH = 0.50 Mole/1.

Similar results were obtained with the lead naphthenate: 1-methoxy-2-propanol system in carbon tetrachloride and cyclohexane solutions as seen in Figure 13 [54].

The data resulting from a tertiary amine (triethylamine): 1-methoxy-2-propanol system in cyclohexane solution are shown in Figure 14 [54]. The downfield shift of the –OH resonance with increasing concentration of catalyst and alcohol. This shift is significantly less than that caused by either of the two metal catalysts. It may be assumed that the hydrogen-bonding type of interaction resulting from the addition of amine to the alcohol is of a lower energy than the oxygen-metal bond in the complex resulting from the addition of either Sn or Pb to the alcohol.

The effect of catalyst combinations on complex formation was also investigated, using solvent-free mixtures of catalyst and alcohol [54]. The shift of the –OH peak was determined on the addition of triethylamine, dibutyltin dilaurate and mixtures of the two catalysts. The data in Table 7 show the shift to be greater for the mixture of the two catalysts than for either catalyst alone, indicating that both hydrogen-bonds and oxygen-metal bonds are formed. The direction of the shifts was opposite to that expected from dilution of the –OH proton. The increased shift appears to correlate with the synergistic effect noted when preparing urethanes with a mixture of these two catalysts.

Efforts were also made to detect complex formation in mixtures of phenyl isocyanate and metal as well as amine catalysts by means of NMR data. However, no significant shifts occurred by the interaction with either of these catalysts.

Since NMR data provided only evidence for alcohol-catalyst complexes but not isocyanate-catalyst complexes, other methods were studied to produce experimental evidence for the latter. The method which was successfully employed for this investigation was determination of molecular weights by means of freezing poing depression [62].

Table 7. Effect of Triethylamine-Dibutyltin Dilaurate Mixture on Chemical Shift of OH Proton in 1-Methoxy-2-Propanol.

Mole Ratio of Alcohol/TEA/DBTDL	Sppm[(a)]
5/0/0	4.22
5/1/0	4.44
5/0/1	4.47
10/1/1	4.94

(a) ppm shift based on TMS

Table 8. Complex Formation in Mixtures of Phenyl Isocyanate and Dibutyltin Dilaurate by Freezing Point Depression of Benzene.

Conc. Phenyl Isocyanate, g/1	Conc. of DBTDL, g/1	Calculated M. W. No Complex (a)	Calculated M. W. 100% Complex (b)	Experimental M.W. (c)	Catalyst in Complex, % $\frac{c-a}{b-a} \times 100$
47.6	0	119	-	119	-
52.4	44.9[1]	187	222	204	49
70.2	157.1[2]	266	392	322	44
46.4	224.4[2]	350	691	482	39

[1] M.W. of DBTDL = 561
[2] M.W. of DBTDL = 575

The molecular weights of the isocyanate and catalyst components were obtained separately at various concentrations in benzene solutions in order to determine if any degree of association existed. The molecular weights of the mixtures were then determined from which the extent of complex formation could be deduced by comparing them with the theoretical values calculated on the basis of (a) no complex formation, and (b) 100% complex formation. The experimental data for mixtures of phenyl isocyanate and dibutyltin dilaurate are shown in Table 8 [62]. These data indicate that as much as 40-50% of the dibutyltin dilaurate molecules are tied up in complexes with phenyl isocyanate. However, it should be noted that in commercially practical urethane reactions, much lower catalyst concentrations are employed.

Similar data, indicating complex formation, were obtained for mixtures of phenyl isocyanate and triethylamine [62].

In order to be certain that these molecular weight values were really evidence of complex formation, a system was selected that would not produce complexes. The system was nonane-DBTDL. Measurements were made to determine whether the experimentally determined molecular weights agree with the calculated values with the assumption of no complex formation. The data in Table 9 [62] show that the experimentally determined molecular weights agree very well with the calculated values for no complex formation. These results lend supporting evidence for complex formation between catalyst and isocyanate and provide additional confidence in the sensitivity of the technique that was utilized.

Rand and McLean [63] employed the mole ratio method [64] in determining complex formation between phenyl isocyanate and metal catalysts (cobalt and manganese naphthenates) in benzene solution. The concentration of the catalyst was held constant and the ratio of phenyl isocyanate was varied from 1:1 to 7:1. Absorbances in the visible spectrum were measured at room temperature in suitable quartz cells with a Cary Model 14 spectrophotometer employing waves lengths of 578 mμ for cobalt and 475 mμ for manganese. Figures 15 and 16 show the plots of the absorbance as a function of the mole ratios. It can be observed in both plots that sharp breaks occurred at the mole

Table 9. Complex Formation in Mixtures of n-Nonane and Dibutyltin Dilaurate by Point Depression of Benzene.

Conc. of n-Nonane, g/1	Conc. of DBTDL, g/1	Calculated M. W. No Complex	Calculated M. W. 100% Complex	Experimental M.W.	Conc. Complex, %
10.2	0	128		127.5	
10.2	241.5	503	597	502	0
41.0	241.5	381	673	383	0

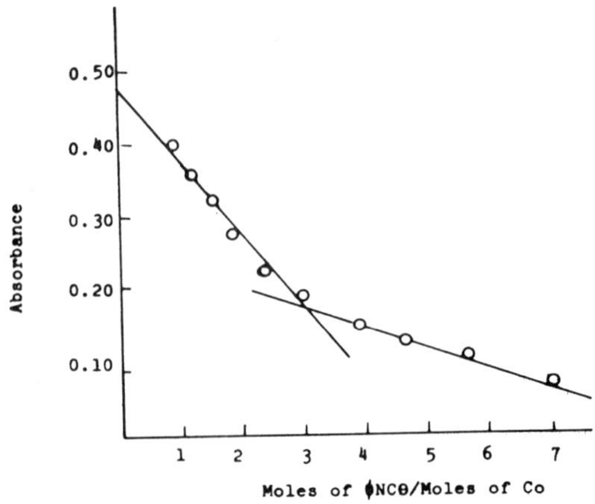

Figure 15. Mole Ratio Plot of Cobalt-Phenyl Isocyanate Complex at 578 mu. (63).

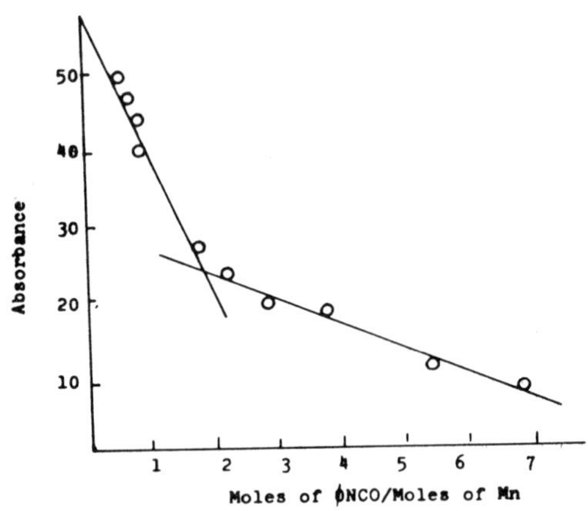

Figure 16. Mole Ratio Plot of Manganese-Phenyl Isocyanate Complex at 475 mul (63).

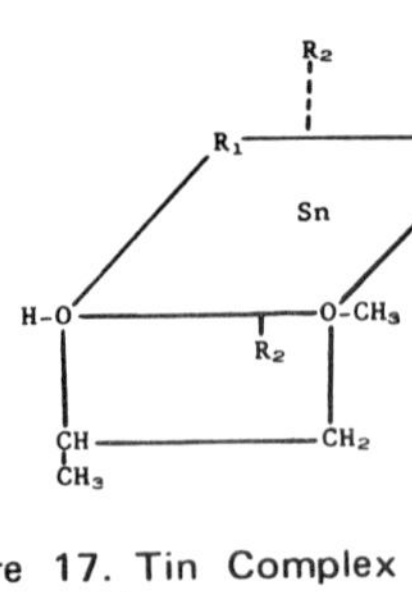

Figure 17. Tin Complex with 1-Methoxy-2-Propanol.

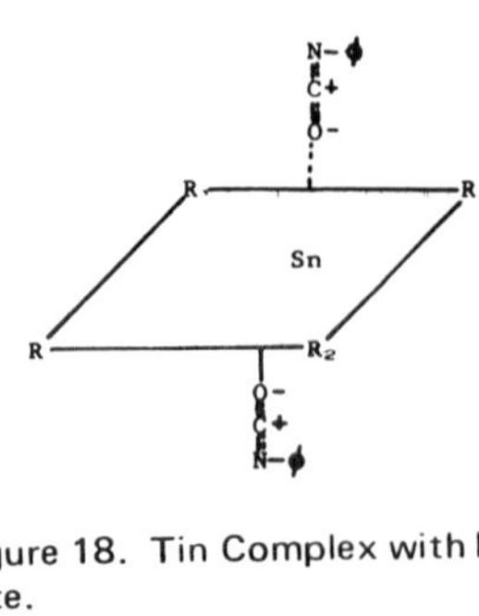

Figure 18. Tin Complex with Phenyl Isocyanate.

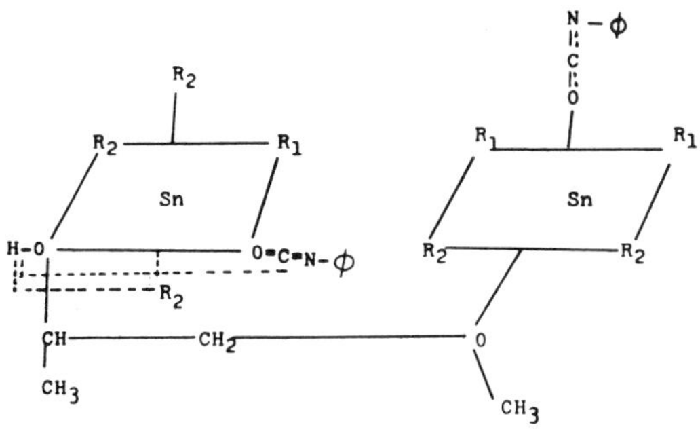

Figure 19. Bridge Complex Between 1-Methoxy-2-Propanol, Phenyl Isocyanate and Tin.

ratio corresponding to the stoichiometry of these complexes, i.e., 3:1 for cobalt and 2:1 for manganese. The nonzero slopes in these graphs were interpreted as meaning that the complexes are dissociated to some extent in solution.

The results of our experimental work with tertiary amines showed clearly complex formation between the amine and the isocyanate as well as between the amine and the hydroxyl component. These findings are in accordance with the mechanism postulated by Baker, et al [65-68].

The results which were obtained by us for complex formation between metal catalysts such as tin and lead compounds, and 1-methoxy-2-propanol and phenyl isocyanate, respectively, were strong experimental confirmation for ternary complex formation first proposed by Britain and Gemeinhardt [40].

The tin complex with 1-methoxy-2-propanol is shown in Figure 17 and the tin complex with phenyl isocyanate in Figure 18. Reaction between the ligands of these complexes presumably takes place by an exchange of the ether oxygen in the complex shown in Figure 17 by an oxygen atom that is part of an isocyanate group as seen in Figure 18. This should result in formation of a bridge compound having the structure shown in Figure 19. Since it is possible for more than one of the tin-alcohol or tin-isocyanate complexes to interact, this could easily account not only for the presence of bridge compounds of the type AB, but also of the type A_2B, A_2B, etc. Urethane formation would then take place by rearrangement of the hydroxyl proton to the nitrogen of the isocyanate group as indicated by the arrow.

Likewise the synergistic action between tin and amine catalysts can be accounted for by the structure in Figure 19 since the isocyanate group can complex with the amine in the manner shown previously.

While the relative speed of complex formation is usually very fast, the rate-determining step, the reaction between the ligands after complex formation has taken place, depends upon the coordination number of the metal ion, the configuration of the complex, and the ionic radius of the ion.

The catalysis in isocyanate reactions has been reviewed extensively by Farkas [69] and Saunders and Frisch [70].

REFERENCES

1. A. W. Hoffmann, Jahresber, 349 (1858)
2. V. E. Shashoua, *J. Am. Chem. Soc. 81,* 3156 (1959)
3. V. E. Shashoua, W. Sweeney and R. F. Tietz, *J. Am. Chem. Soc. 82,* 866 (1960)
4. J. S. Blair and G. E. P. Smith, Jr. *J. Am. Chem. Soc. 56,* 907 (1934)
5. W. Frentzel, Ber. *21,* 411 (1888)
6. J. M. Lyons and R. H. Thompson, *J. Chem. Soc.* 1971 (1950)
7. L. C. Raiford and H. B. Freyermuth, *J. Org. Chem. 8,* 230 (1943)
8. R. G. Arnold, J. A. Nelson and J. J. Verbanc, *Chem. Revs. 57,* 47 (1957)
9. O. Bayer, Bios Final Report No. 719, "Interview with Professor Otto Bayer," July 1946.
10. A. W. Hoffman, Ber. *18,* 764 (1885)
11. A. Michael, Ber. *38,* 22 (1905)
12. W. J. Balon, U. S. Pat. 2,801,244 (to duPont) (1957)
13. H. Havekoss, "Polymerization of Diisocyanates", O. P. B. Report 73894, Frames 4709-18, April 10, 1942
14. A. W. Hoffmann, Ber. *3,* 761 (1870)
15. T. M. Laakso and D. D. Reynolds, *J. Am. Chem. Soc. 79,* 5717 (1957)
16. H. Kleiner, H. Havekoss and F. V. Spulak (to Farbenfabriken Bayer), Ger. Pat. 872, 618 (1941)
17. I. C. Kogon, *J. Org. Chem. 23,* 1594 (1958)
18. I. I. Jones and N. G. Savill, *J. Chem. Soc.* 4392 (1957)
19. B. D. Beitchman, Rubber Age *98,* No. 2,65 (1966)
20. J. Burkus, (to U. S. Rubber Co.) U. S. Pat. 2,979,485 (1961)

20-A L. Nicholas and G. T. Gmitter, *J. Cellular Plastics 1,* No. 1, 85 (1965)

21. G. Nischk (to Farbenfabriken Bayer), Ger. Pat. 924, 751 (1955)
22. R. Stolle, Ber. *41,* 1125 (1908)
23. E. Dyer and G. E. Newborn, *J. Am. Chem. Soc. 80,* 5495 (1958)
24. T. W. Campbell and J. J. Monagle, *J. Am. Chem. Soc. 84,* 1492 (1962)
25. W. Neumann and P. Fischer, *Angew, Chem. 74,* 801 (1962)
26. T. W. Campbell, J. J. Monagle and V. S. Foldi, *J. Am. Chem. Soc. 84,* 3673 (1962)
27. J. J. Monagle, T. W. Campbell and H. F. McShane, Jr. *J. Am. Chem. Soc. 84,* 4288 (1962)
28. W. J. Balon, (to duPont), U. S. Pat. 2,853,518 (1958)
29. W. Neumann, J. Peter, H. Holtschmidt and W. Kallert, Preprints, 4th Rubber Technol. Conf. London, May 22-25, 1962
30. W. C. Darr, P. G. Gemeinhardt and J. H. Saunders, *J. Cellular Plastics 2,* 266 (1966)
31. G. G. Greth, R. G. Smith and G. O. Rudkin *J. Cellular Plastics, 1,* 159 (1965)
32. B. G. Alzner and K. C. Frisch, *Ind. Eng. Chem. 51,* 715 (1959)
33. A. Farkas, G. A. Mills, W. E. Erner, and J. B. Maerker, *Ind. Eng. Chem. 51,* 1299 (1959)
34. F. Mort, J. Oil & Colour Chem. Associate *45,* 95 (1962)
35. H. W. Wolfe, Jr., E. I. duPont de Nemours & Co., Elastomers Foam Bulletin "Tin Catalyst Activity in Urethane Foam," Feb. 24, 1961.
36. A. F. Rogers, "Catalysis of the Reactions Involved in Making Flexible Urethane Foams," Polymer Conference on Cellular Plastics, Wayne State Univ., May 1966.
37. J. W. Baker and J. B. Holdsworth, *J. Chem. Soc.* 713 (1947)
38. V. S. Pshezetsky, I. Massouh and V. A. Kabanov, J. Polymer Sci., Part C, *22,* 309 (1968)
39. F. Hostettler and E. F. Cox, *Ind. Eng. Chem. 52,* 609 (1960)
40. J. W. Britain and P. G. Gemeinhardt, *J. Appl. Polymer Sci. 4,* 207 (1960)
41. T. E. Lipatova and Y. N. Nizelskii, Dokl. Akad. Nauk SSSR *175,* No. 1, 128 (1967)
42. I. R. Fischer, Tetrahedron, *19,* Suppl. 1, 97 (1963)
43. V. I. Avdeev and L. I. Zakharov Teoretich. i eksp. khim. *2,* 5(1966)
44. S. L. Reegen and K. C. Frisch, *J. Polymer Sci.* A-1, *4,* 2321 (1966)
45. H. Okada and Y. Iwakura, Makromol, *Chem. 64,* 91 (1963)
46. M. Sato, *J. Org. Chem. 27,* 819 (1962)
47. I. C. Kogon, *J. Org. Chem. 26,* 3004 (1961)
48. G. Gutnikov and G. N. Schenk, *Anal, Chem. 34,* 1316 (1962)

49. H. A. Smith, *J. Polymer Sci.* A-1, *6,* 1299 (1968)
50. J. W. Baker, J. Gaunt and M. M. Davies, *J. Chem. Soc.* 9 (1949)
51. A. Farkas and K. G. Flynn, *J. Am. Chem. Soc. 82,* 642 ((1960)
52. L. Rand, B. Thir, S. L. Reegen and K. C. Frisch, *J. Appl. Polymer Sci. 9,* 1787 (1965)
53. K. C. Frisch, S. L. Reegen and B. Thir, *J. Polymer Sci.* Part C, *16,* 2191 (1967)
54. K. C. Frisch, S. L. Reegen, W. V. Floutz and J. P. Oliver, *J. Polymer Sci. 5,* 35 (1967)
55. H. A. Smith, *J. Appl. Polymer Sci. 7,* 85 (1963)
56. M. Pestemer and D. Lauerer, *Angew, Chem. 72,* 612 (1960)
57. A. Farkas and P. F. Strohm, *I & EC Fundamentals 4,* No. 1, 32 (1965)
58. D. C. Bradley, E. V. Caldwell and W. Wardlaw, *J. Chem. Soc.* 3039 (1957)
59. A. J. Bloodsworth and A. G. Davis, Proc. Chem. Soc. 264 (1963)
60. E. Dyer and R. B. Pinkerton, *J. Appl. Polymer Sci. 9,* 1713 (1965)
61. S. G. Entelis, O. V. Nesterov and R. P. Tiger, *J. Cellular Plastics, 3,* 360 (1967)
62. S. L. Reegen and K. C. Frisch, *J. Polymer Sci.* Part A-1, *8,* 2883 (1970)
63. L. Rand and J. A. McLean, Jr., *J. Appl. Polymer Sci. 12,* 467 (1968)
64. J. H. Yoe and A. L. Jones, *Ind. Eng. Chem. Anal. Ed. 16,* 111 (1944)
65. J. W. Baker and J. Gaunt, *J. Chem. Soc.* 9; *ibid.* 19 (1949)
66. J. W. Baker and J. Gaunt, *J. Chem. Soc.* 24; *ibid,* 27 (1949)
67. J. W. Baker, M. M. Davies and J. Gaunt, *ibid,* 24 (1949)
68. J. W. Baker and D. N. Bailey, *ibid,* 4649; 4652; 4463 (1957)
69. A. Farkas and G. A. Mills, "Catalytic Effects in Isocyanate Reactions" in "Advances in Catalysis" Vol. 13, Academic Press, New York (1962)
70. J. H. Saunders and K. C. Frisch, Chapter IV, "Kinetics and Catalysis of the Isocyanate Reactions" in "Polyurethanes," Part I, Wiley-Interscience, New York (1962)

S. L. REEGEN and K. C. FRISCH, *Polymer Institure, University of Detroit*

CHAPTER 2

UNCONVENTIONAL CHEMISTRY OF ISOCYANATES

INTRODUCTION

The reaction of alcohols with isocyanates to produce carbamates (urethanes) was discovered by Wurtz in 1848. O. Bayer's application of this reaction to polyfunctional alcohols and isocyanates in 1937 created the polyurethane industry. Today only three decades later, we are witnessing the enormous economical success of polyurethanes. This success in part is related to the versatility of the urethane systems which allows fabrication of polymeric materials ranging from soft elastoplastics to rigid foams.

A wide variety of inexpensive raw materials are commercially available, and the current trend in the industry is toward broader use of existing materials. Nevertheless, a challenge is presented to chemists to look in new directions, and to further widen the scope of application of urethane chemicals. Especially the chemical reactivity of isocyanates, coupled with a relatively low price, makes these compounds the logical choice for the design of new unique polymer systems.

Already polyisocyanurates and heterocyclic polymers derived from diisocyanates are on the market, and it is the purpose of this article to focus attention to new uses of diisocyanates. Most of the presented matter is selected from the recent literature, but the discussion of unconventional isocyanate chemistry in some detail may provide food for thought for the inventive chemist.

I have subdivided this lecture into four topics; new trends in urethane chemistry, polycondensation, cycloaddition and insertion reactions of isocyanates. Especially the cycloaddition and insertion reactions of isocyanates are not widely known reactions, but it is indicated that they will find utility in the plastics industry in years to come. A review of this type of chemistry has appeared recently [1].

A. NEW TRENDS IN URETHANE CHEMISTRY

In order to widen the scope of urethane applications, it is necessary to solve problems related to thermal and color stability. Addition polymers, such as polyurethanes, have the advantage that no by-products are formed in their production. However, the thermal reversibility of the addition reaction presents a problem. While nothing could be done to improve the thermal stability of conventional polyurethanes, a combination of polyaddition, followed by polycondensation can be used to generate thermally more stable polymers. This approach will be discussed in Part B. A second approach to thermally more stable polymers involves cycloaddition reactions of isocyanates, and examples are presented in Part C.

Another shortcoming of aromatic diisocyanate based polyurethanes is their discoloration upon exposure to air and ultraviolet light. In contrast, polyurethanes based on aliphatic diisocyanates are color stable, but low priced monomers are not yet available. The aliphatic diisocyanates, which are presently produced in commercial or development quantities are shown in Table 1. All of the depicted materials have certain drawbacks. The economically attractive HDI can only be used in prepolymer or biuret form because of its high toxicity. Reduction of MDA to $H_{12}MDA$ requires high pressure and nobel metal catalysis; also, it is absolutely essential for color stability that no residual aromaticity remains in the product.

Furthermore, mixtures of stereoisomers are produced in the reduction of aromatic diisocyanates. For example, $H_{12}MDI$ is a mixture of three isomers. The commercially available xylidene diisocyanate is also a mixture of the m- and p-isomer. The rate of reaction of aliphatic diisocyanates is considerably slower than aromatic diisocyanates and cycloaliphatic secondary isocyanate groups are slower reacting than linear primary aliphatic diisocyanates. Of

Table 1. Aliphatic Diisocyanates.

$OCN(CH_2)_6NCO$

HDI

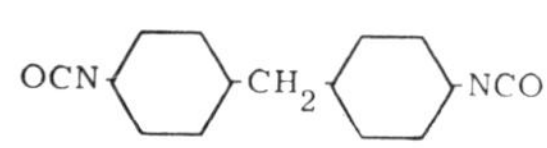

$H_{12}MDI$

$OCNCH_2C(CH_3)_2CH_2CH(CH_3)CH_2CH_2NCO$

TMDI

IPDI

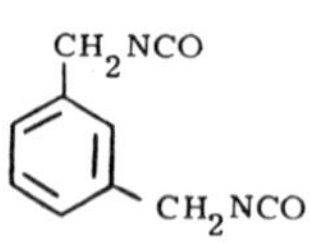

TAKENATE

Table 2. Carbonyl and Sulfonyl Diisocyanates.

OCNCONCO

$OCNSO_2NCO$

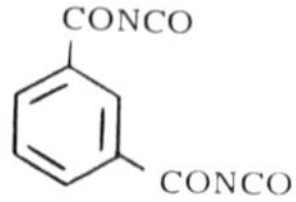

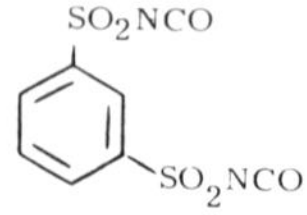

course, the slower reaction rate requires higher concentrations of catalysts, and it is well known that the reverse reaction, i.e., degradation of polyurethanes, is catalyzed by bases.

A different approach to the discoloration problem requires the use of diisocyanates, in which the isocyanate group is attached to a carbonyl or a sulfonyl group. In this case the rate of reaction of the isocyanate group is considerably enhanced, and the resulting polymers are expected to be color stable. Several attractive monomers of this kind are listed in Table 2.

We have synthesized m-phenylenedisulfonyl diisocyanate by direct phosgenation of the corresponding disulfonamide in the presence of a catalytic amount of n-butyl isocyanate. Polyurethanes, derived from sulfonyl diisocyanates are color stable, but the acidic proton of the sulfonyl carbamate group induces base solubility. This property had been used to prepare ion exchange resins from 4-isocyanatobenzenesulfonyl isocyanate and starch [2]. However, in an all-purpose coating, base solubility is not desirable. Isophthaloyl diisocyanate has recently been synthesized [3]; however, to the best of my knowledge, no polyurethanes based on this monomer have been prepared.

The correct handling of these highly reactive monomers is very important because acyl and sulfonyl isocyanates react instantaneously with atmospheric moisture.

Economical consideration will eventually be decisive and the least expensive monomer will be used in formulations where color stability is essential.

Another important problem related to the use of polyurethanes is the moisture sensitivity of the isocyanate component which often causes problems in the field. These problems are magnified in the area of cast elastomers, where a quality polymer is produced often by unqualified labor. The solution to this problem would be a one-component system which is polymerized by the application of heat or by the addition of a catalyst.

In coating applications presently so-called blocked isocyanates are used. The volatile blocking agent is released upon heating and the generated diisocyanates undergo rapid polymerization with the present polyester or polyether

Scheme 1. Homopolymerization of Isocyanatophenols.

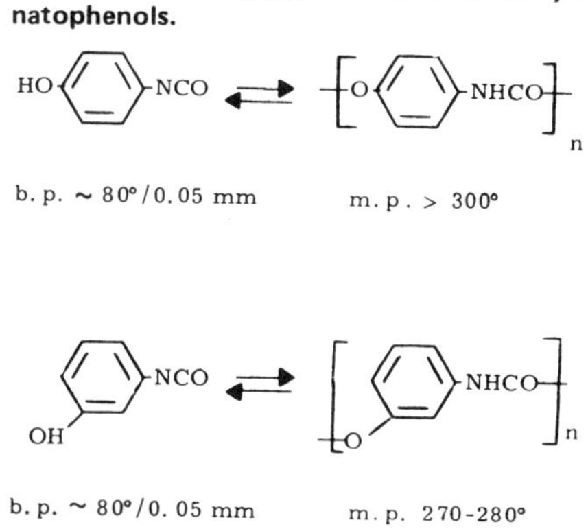

diols. Drawbacks of this method are the limit in thickness which can be obtained, the expense of the blocking agent and the necessity to use well ventilated working areas. The ideal system would involve reactive monomers, stable at room temperature, and addition of heat or a catalyst would trigger polymerization. For example, unsubstituted isocyanatophenols are stable in solution. Addition of a base catalyst triggers polymerization to the corresponding polyurethane (Scheme 1) [4].

Here we are dealing with a model system; however, it could be visualized that substituents attached in ortho position to the functional groups could provide sufficient steric hindrance to have a stable difunctional monomer which can be homopolymerized by base or metal organic catalysis. No volatile by-products are generated, and heat curing may provide an increase in thermal stability.

Reaction of diisocyanates with glycidol produces isocyanate terminated carbamates, which upon heating in the presence of suitable catalysts rearrange to 4-hydroxymethyl-2-oxazolidones. Subsequent reaction of the generated hydroxy group with the present isocyanate yields a polyurethane [5, 6] (Equation 1).

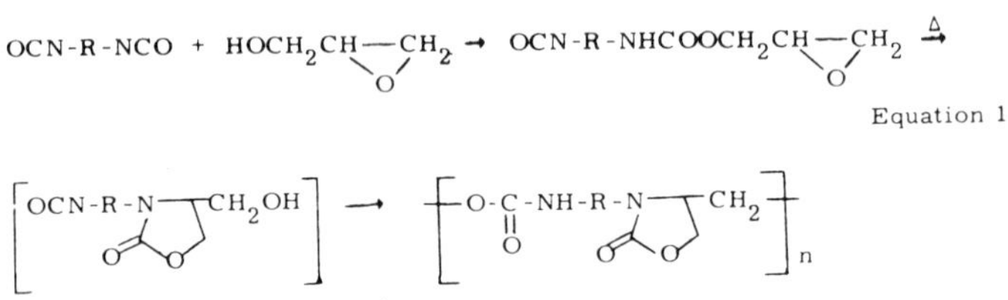

Equation 1

The blocking agent becomes part of the polymer system in the above approach, however, shelf life of the isocyanate terminated intermediate, especially in the presence of a catalyst, presents a problem. Chemical innovation is still required to design workable one component systems.

B. POLYCONDENSATION

Addition of isocyanates to active hydrogen containing compounds creates bonds which are thermally labile. For example, polyurethanes dissociate above 150°C. In order to obtain polymeric materials which are stable at more elevated temperature, condensation-type polymerization is required. Interestingly, attempts have been made to synthesize condensation polymers from diisocyanates and suitable substrates. Sorensen and Campbell [7] reported the synthesis of polydecamethylenesebacamide from decamethylene diisocyanate and sebacic adic (Equation 2).

$$OCN(CH_2)_{10}NCO \; + \; HOOC(CH_2)_8COOH \longrightarrow$$

$$-\!\!\left[NH(CH_2)_{10}NHCO(CH_2)_8CO\right]\!\!-_n \; + \; 2\,CO_2$$

Equation 2

However, since polyamides are more readily prepared from the corresponding diamines and dicarboxylic acids, this reaction is of no commercial significance.

In contrast, polyimides are considerably more stable than polyamides, and in the preparation of polyimide foams from diisocyanates and dianhydrides the generated carbon dioxide can be used as the blowing agent (Equation 3).

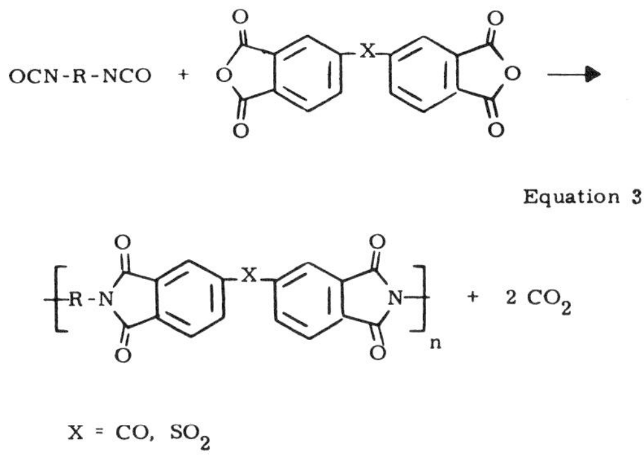

Equation 3

It has been demonstrated in our laboratory that polyimide foams can be obtained if the reaction of the diisocyanate with the dianhydride is being conducted in dimethylsulfoxide in the presence of a catalytic amount of an active hydrogen containing compound [8].

The obtained open-celled foams have good physical properties and excellent thermal stability and fire resistance. Samples held at 450°F (232°C) in air for 14 days suffered weight loss of less than 2% with no loss in compression strength. A 6% weight loss was measured in 10 minutes at 842°F (450°C). A flame spread rating of 10 was obtained in the ASTM E-84 tunnel test.

A variety of temperature stable heterocyclic polymers have been synthesized by reacting diisocyanates with tetrafunctional monomers. For example, reaction of diisocyanates with the diphenylester of resorcinol dicarboxylic acid produces in the first step the linear carbamate polymers, which are cyclized to polybenzoxazindiones upon subsequent cyclization with elimination of phenol (Equation 4) [9].

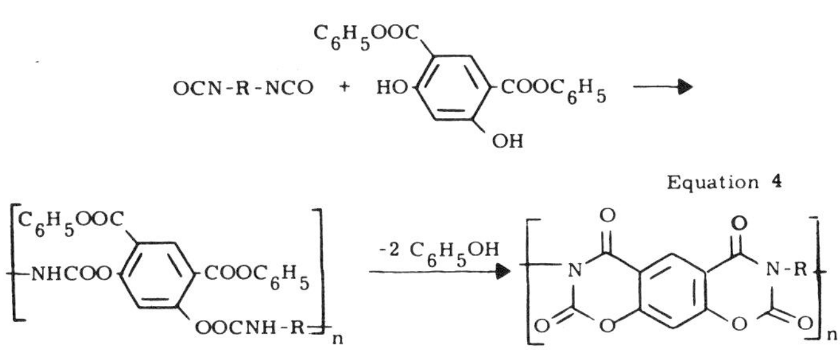

Equation 4

The reaction is conducted in DMSO as solvent in the presence of a tertiary amine catalyst. The cyclization occurs at approximately 90°C. The resulting films and fibers are thermally stable up to 300°C.

From diaminodicarboxylic acids and diisocyanates the thermally stable polyquinazolinediones are obtained (Equation 5) [10]. Again, the diphenyl

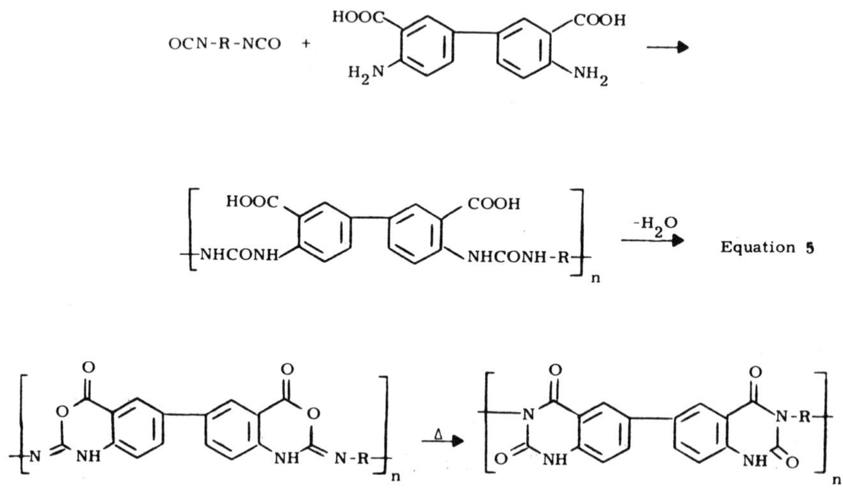

Equation 5

esters can be used instead of the free acids, and this reaction requires a thermal rearrangement of the initially formed oxazinone polymers as evidenced by DSC analysis. The TGA analysis of the polyquinazolinediones indicates thermal stability up to 500°C.

Of course, a wide variety of heterocyclic addition/condensation polymers are visualized, and eventually the economically most attractive candidates will be used.

C. CYCLOADDITION

The cycloaddition reactions of heterocumulenes are well known and already A. W. von Hofman in 1862 has observed that aromatic isocyanates undergo dimerization and trimerization reactions. In 1960 Shashoua and coworkers [11]

Scheme 2. Isocyanate Oligomers.

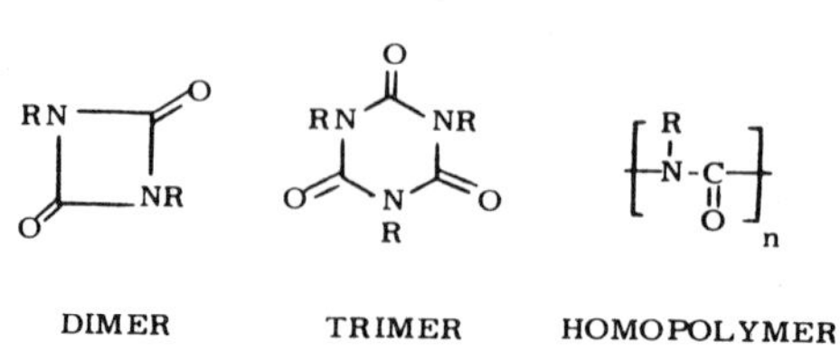

demonstrated that homopolymers can be obtained by anionic polymerization of alkyl and aryl isocyanates, provided the reaction is conducted at low temperatures. The various types of oligomerization reactions of isocyanates are shown in Scheme II.

The formation of trimeric and polymeric species requires usually catalysis. In contrast, dimerization of certain symmetrical diisocyanates can occur in the absence of catalysts, and the rate of dimerization is faster in the solid state than in the liquid phase. The dimeric species can be converted to the monomers upon heating, and therefore polydimers are of no significance. In contrast, homopolymers and cyclic trimers are considerably more stable toward heat.

While the base catalyzed homopolymerization of mono isocyanates afford linear polyamides, heterocyclic polymers can be obtained from aliphatic 1, 2 and 1, 3-diisocyanates. For example, cyclopolymerization of ethylene diisocyanates yields polyimidazolidones (Equation 6) [12].

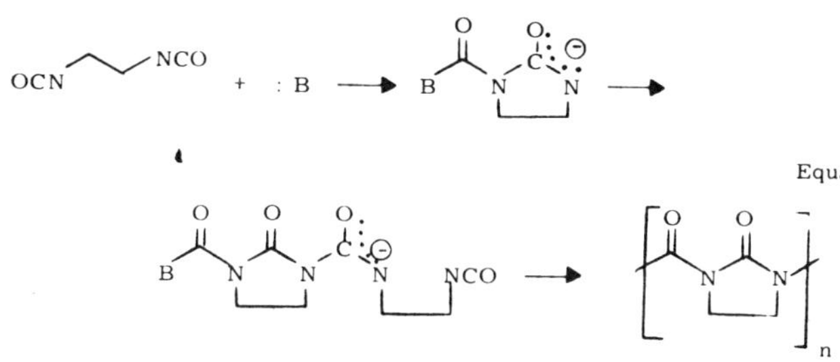

Equation 6

Cyclopolymeric ureas are also obtained from vinyl isocyanate. The first step in this reaction is the homopolymerization of the isocyanate group, followed by cycloaddition of the vinyl group (Equation 7) [13].

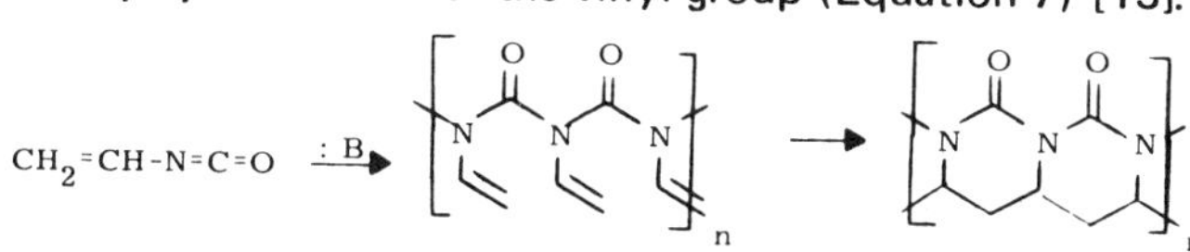

Equation 7

The cyclotrimerization of di and polyisocyanates has received considerable attention lately because heat resistant rigid foams can be produced in this reaction [14]. The physical properties of the prepared isocyanurate foams were quite good and particularly noteworthy is the high temperature performance.

The above outlined examples of cycloaddition reactions involve cycloaddition reactions of isocyanate groups. Of course, addition of isocyanate groups to an enormous variety of double bonded substrates is well known. A general description of this reaction is shown in Scheme III [15]. From this general description it is clear that a variety of products can be formed depending upon reaction conditions and substrates. The 1:1 adducts are either polar linear species or four-membered ring heterocycles, and the 2:1 or 1:2 adducts are six-membered ring heterocycles. At elevated temperature the exchange

Scheme 3. Polar Cycloaddition of Isocyanates to Double Bond Containing Substrates

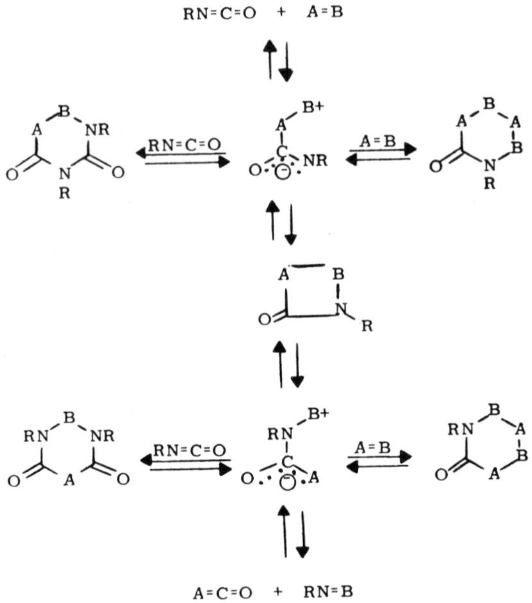

Scheme 4. Cycloaddition of Isocyanates to Carbodiimides.

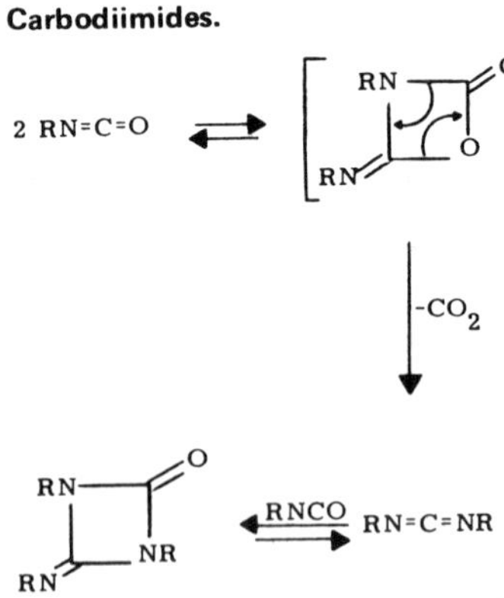

reactions predominate, and if one gaseous or low boiling component is removed from the reaction equilibria, exchange products can be formed in virtually quantitative yield.

An excellent example of the exchange reaction is the formation of carbodiimides from isocyanates. If carbodiimide linkages are generated in the presence of isocyanates, addition of isocynate groups to the generated carbodiimide occurs [16]. At elevated temperature reversal of this reaction is observed, and therefore carbodiimides can be used to block isocyanate groups (Scheme IV). One example of utilization of this reaction is liquid

MDI. MDI is a low melting solid material and for ease of handling we are marketing a liquid version which is based on the outlined chemistry.

The formation of carbodiimides from isocyanates requires base catalysis, but trimerization completes with carbodiimide formation. However, carbodiimides are generated exclusively if phospholene oxide catalysts are used. The reaction mechanism involves a series of polar 1,2-cycloaddition reactions. Polycarbodiimides have been prepared from diisocyanates using a phospholene oxide catalyst (Equation 8) [17].

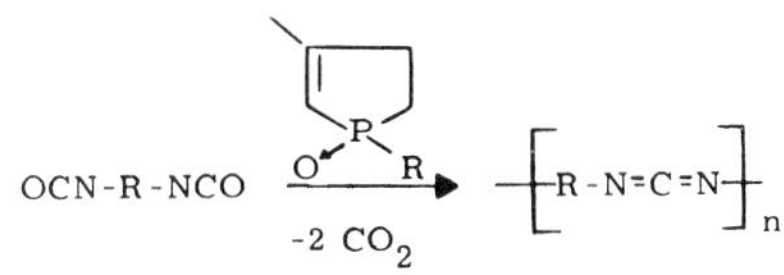

Equation 8

While the [2 + 2] cycloaddition reaction of isocyanates to activated C=C and C=N bonds usually yields four-membered ring 1:1 adducts or six-membered ring 2:1 adducts, exchange reactions occur if the substrates have C=O, P=N or S=O double bonds.

The cycloadducts are often thermolabile, however, if hydrogen atoms are adjacent to the carbonyl group in the four-membered ring adduct rearrangement can occur. For example, addition of isocyanates to enamines having hydrogen atoms in the β -position, produces the corresponding thermally stable linear amides (Scheme V).

Scheme 5. Cycloaddition of Isocyanates to Enamines.

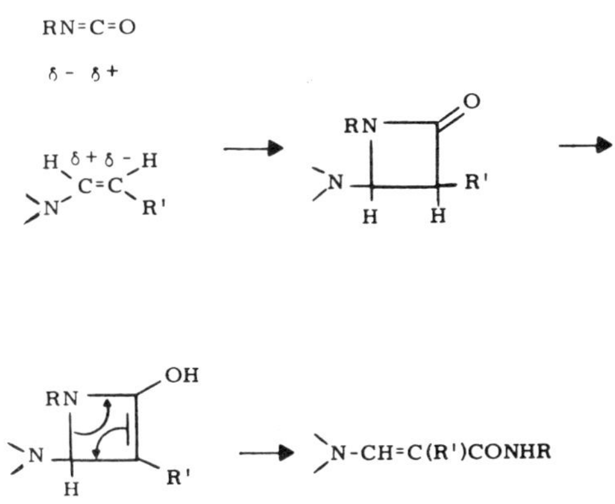

Polymer examples utilizing this reaction to produce polyamides from diisocyanates and enamines have recently been reported [18, 19]. Thus, reaction of MDI with an enamine of a cyclic ketone has produced a polyamide after hydrolysis of the enamine function (Equation 9) [18]. Amide bonds

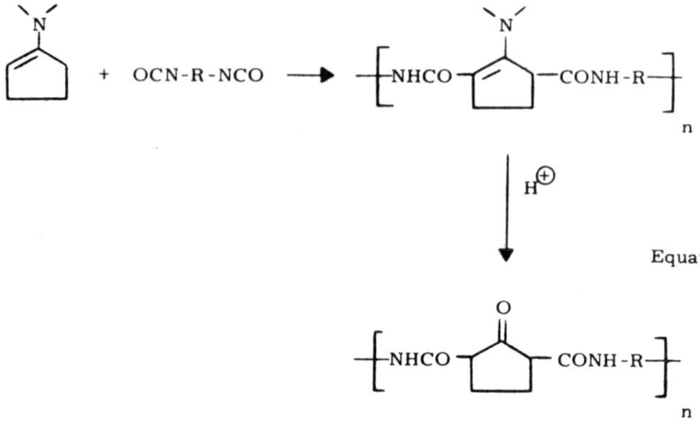

Equation 9

can also be obtained in a similar manner from isocyanates and C=N double bond containing substrates, capable of hydrogen shift. A wide variety of cyclic imines, amidines, oxazolidines, etc. undergo this type of reaction [20].

Isocyanates also react as dipolarophiles in 1, 3 and 1, 4-cycloaddition reactions (Scheme VI). For example, reaction of diisocyanates with diepoxides in the presence of a base catalyst produces polyoxazolidinones which are thermally quite stable (Equation 10) [21]. Since the monomers are readily available,

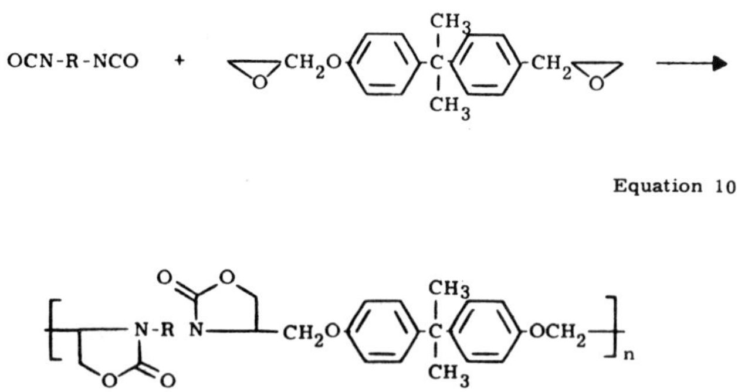

Equation 10

utility of these materials in the plastics industry is anticipated.

The same 1, 3-dipolar substrates can also be generated from the corresponding bis-ethylene carbonates.

An example of a 1, 4-dipolar cycloaddition reaction involving isocyanates as dipolarophiles is the reaction with an iminoketene generated *in situ* (Scheme VII). The resulting cycloadducts are thermally quite stable, and in the case of difunctional intermediates polyquinazolinediones are formed (see Part B). As by-products only carbon dioxide or nitrogen gas are produced. Also a [2 + 2 + 2] cycloaddition sequence, involving C=N double bonded substrates and two isocyanate groups, can give rise to six-membered ring heterocycles, which are thermally and chemically stable [20]

Scheme 6. 1,3, and 1,4–Cycloaddition Reactions.

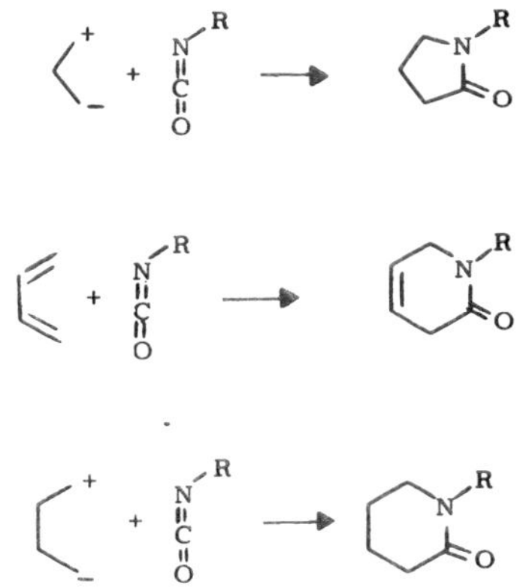

Scheme 7. Cycloaddition of Isocyanates to Iminoketenes.

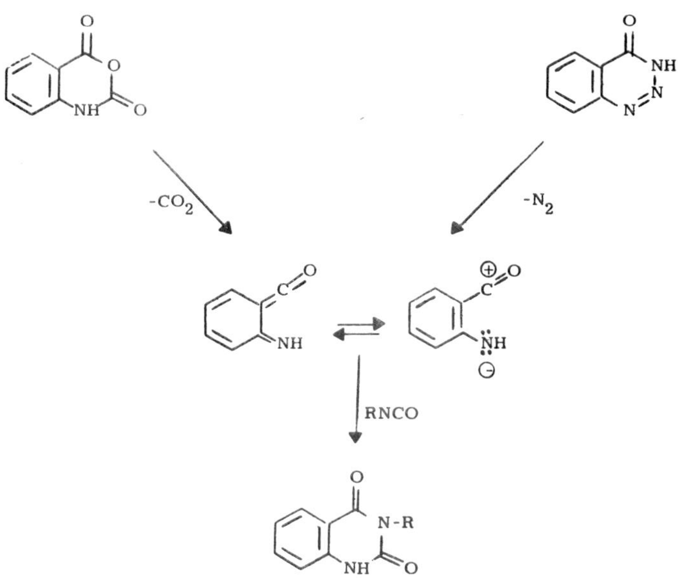

Isocyanates, having a thioacyl or an imidoyl group attached to the isocyanate group are seldom stable in the monomeric form. For example, attempts to generate arylimidoyl isocyanates by a variety of methods resulted in the formation of a heterocyclic system formed by an intramolecules [2 + 4]cycloaddition reaction (Equation 11) [22].

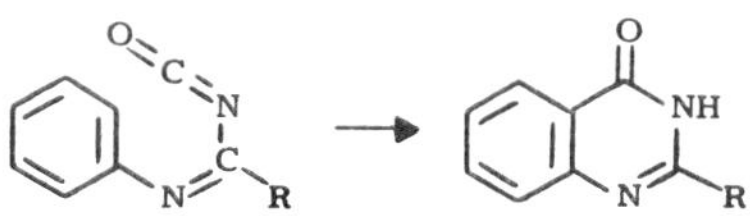

Equation 11

Scheme 8. Cycloaddition Reaction of 2-Isocyanatopyridine.

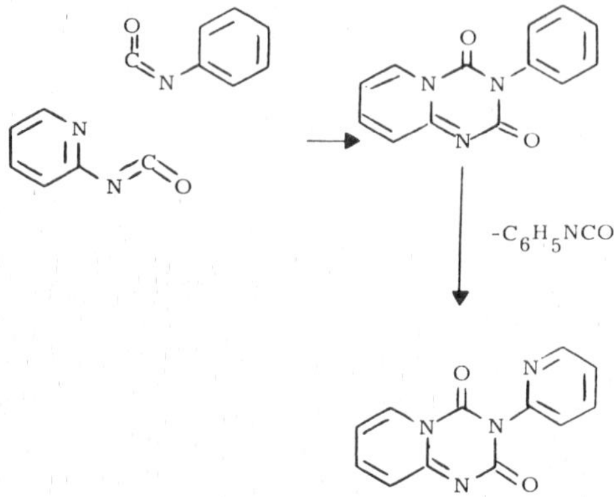

The instability of heterocyclic isocyanates can be explained by a rapid [2 + 4] cycloaddition reaction. For example, monomeric 2-isocyanatopyridine can be generated by a variety of methods, and the monomer can be trapped with phenyl isocyanate to produce a 1:1 adduct [23]. Upon heating phenyl isocyanate is generated and the 2-isocyanatopyridine cyclodimer is obtained (Scheme VIII). Several other heterocyclic isocyanates can be generated *in situ* and trapped as cycloadducts or as cyclodimers [24].

Application of this reaction to difunctional compounds could lead to a wide variety of new heterocyclic polymers. An interesting example of a series of cycloaddition reactions has been observed in the reaction of aryl isocyanates with N, N-dimethylformamide, a commonly used solvent in polymer chemistry [25]. The initial reaction is a [2 + 2] cycloaddition reaction with elimination of carbon dioxide to form an amidine. The amidine reacts with isocyanate to produce a 1, 4-dipol which is intercepted by another isocyanate molecule to form a heterocyclic molecule by a [2 + 4] polar cycloaddition sequence. The heterocyclic compound reacts further with isocyanate by insertion into a carbon to nitrogen bond (see next part) to produce a heterocyclic urea derivative. The urea derivative reacts with two molecules of the isocyanate to produce a spiro compound which is thermally very stable (Scheme IX).

Of course, numerous examples of cycloaddition reactions of monoisocyanates are listed in the literature [1, 20], and utilization of this type of chemistry in the preparation of new interesting polymer systems is anticipated.

D. INSERTION REACTIONS

The nucleophilic reactions of isocyanates involve addition of a nucleophile to the electrophilic carbon atom of the isocyanato group. On the other hand, isocyanates can also react as nucleophiles. The electronic configuration of a single bonded substrate would determine the initial site of reaction. For example, in the reaction of an isocyanate with a polar single bonded substrate A-B the

Scheme 9. Reaction of Isocyanates with DMF.

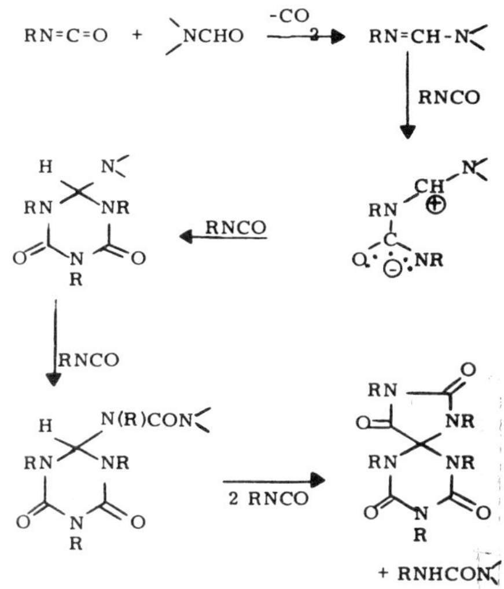

electron pair on the nitrogen or oxygen of the cumulative double bonds could attach the electrophilic center A, or the nucleophilic center B becomes initially attached to the electrophilic center carbon in the N=C=O group. If the reaction is accompanied by a 1, 3-rearrangement, insertion of one of the double bonds of the isocyanate group into the polar single bond has occurred (Equation 12).

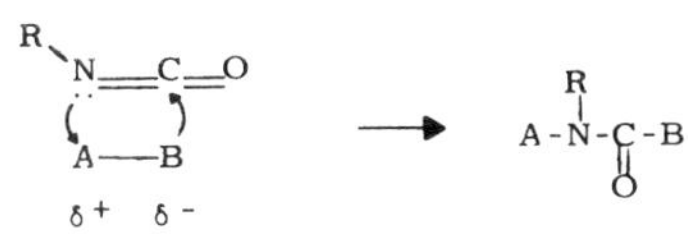

Equation 12

This reaction may in some cases proceed via a four centered transition state, very much like the Wittig reaction of phosphorusylides. In Table 3 a variety of single bonded substrates are listed which are known to form 1:1 adducts with isocyanates.

An example, involving diisocyanates is the formation of a silicon containing triisocyanate from hexamethylcyclotrisilazane and MDI [26] (Equation 13).

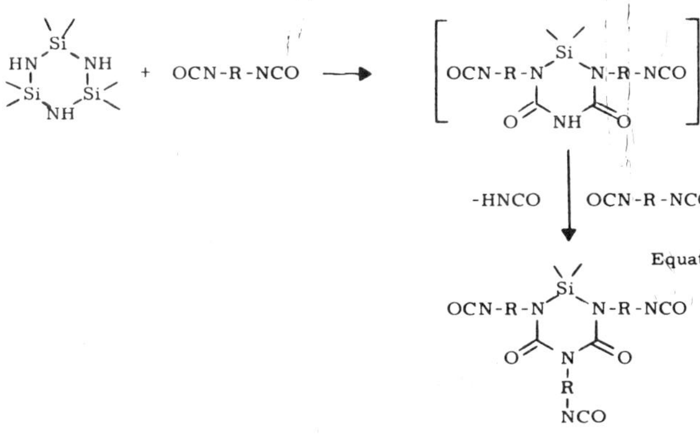

Equation 13

This reaction is initiated by a [2 + 2] cycloaddition across the Si-N bond followed by ring enlargement and further reaction with isocyanate to produce the intermediate diisocyanate. Further insertion of isocyanate and elimination of cyanic acid produces the finally isolated product.

Often the generated 1:1 adducts are more reactive than the starting material because a nitrogen to substrate bond is formed. Insertion of a second isocyanate molecule produces a 2:1 adduct and this process can be repeated to produce cyclic trimers with ejection of the substrate A-B. In these cases A-B is acting as a trimerization catalyst. If a stepwise selective insertion can be accomplished trisubstituted isocyanurates can be produced cleanly; however, due to the reversibility of the outlined processes, scrambling can occur. Polymers, generated by this insertion process ought to be formed quite easily, but the hydrolytic stability of the newly formed substrate to nitrogen bonds may be the limiting factor.

In summary, isocyanates are remarkably reactive organic molecules, and their utility in the formation of new polymer systems is clearly indicated. In the next decade, tailor-made polymers for many new applications will be synthesized and diisocyanates will play an important role in this endeavor.

Table 3. Insertion Reactions of Isocyanates

Group of Periodic System	Element	Bonds Involved			
II	Magnesium	Mg-C			
	Zinc	Zn-C			
	Cadmium	Cd-C			
	Mercury			Hg-O	
III	Boron	B-C	B-N	B-O	B-C1
	Aluminum	A1-C	A1-N	A1-S	
IV	Carbon		C-N	C-O	
				C-S	
	Silicon		Si-N	Si-O	
			Si-P		
	Germanium		Ge-N		
	Tin		Sn-N	Sn-O	
			Sn-P	Sn-S	
	Lead		Pb-N	Pb-O	
V	Phosphorus		P-N		
	Arsenic		As-N		
VI	Sulfur		S-N		S-C1

REFERENCES

1. H. Ulrich, "Cycloaddition Reactions of Heterocumulenes," Academic Press, New York, 1967.
2. R. W. Rousseau, C. D. Callihan and W. H. Daly, *Macromolecules 2,* 502 (1969).
3. O. Tsuge, T. Itoh and M. Tashiro, *Tetrahedron 24,* 2583 (1968).
4. H. Ulrich, U. S. Patent 3,488,376 (1970).
5. M. L. Kaufmann, Ger. Offen. 1,917,166 (1969).
6. W. J. Farrissey, Jr. and A. M. Nashu, *J. Heteroc. Chem. 7,* 331 (1970).
7. W. R. Sorenson and T. W. Campbell, "Preparative Methods of Polymer Chemistry," 2nd Edit., p. 102, Wiley, New York 1968.
8. W. J. Farrissey, Jr., J. S. Rose and P. S. Carleton, *J. Appl. Polymer Sci. 14,* 1093 (1970).
9. L. Bottenbruch, *Angew. Chem. 81,* 911 (1969).
10. N. Yoda, R. Nakanishi, M. Kurihara, Y. Bamba, S. Tohyama and K. Ikeda, *Polymer Letters 4,* 11 (1966).
11. V. E. Shashoua, W. Sweeny and R. F. Tietz, *J. Am. Chem. Soc. 82,* 866 (1960).
12. Y. Iwakura, K. Uno and K. Ichikawa, *J. Poly. Sci., A2,* 3387 (1964); C. King, *J. Am. Chem. Soc. 73,* 2959 (1951).
13. C. G. Overberger, S. Ozaki and H. Mukamal, *J. Poly. Sci. B2,* 627 (1964).
14. G. W. Ball, G. A. Haggis, R. Hurd and J. F. Wood, *J. Cell. Plastics 4,* 248 (1968).
15. H. Ulrich, *Acc. Chem. Res. 2,* 186 (1969).
16. W. J. Farrissey, Jr., R. J. Ricciardi and A. A. R. Sayigh, *J. Org. Chem. 33,* 1913 (1968).
17. T. W. Campbell and K. C. Smeltz, *J. Org. Chem. 28,* 2069 (1963).
18. W. H. Daly, Polymer Preprints of the 152nd Meeting of the American Chemical Society, p. 569, New York, 1966.
19. K. Harada, Y. Mizoe, J. Furukawa, and S. Yamashita, *Makrom. Chem. 132,* 295 (1970).
20. H. Ulrich and R. Richter, "Newer Methods of Preparative Organic Chemistry," Volume 6, Academic Press, New York, 1970.
21. S. R. Sandler, F. Berg and G. Kitazawa, *J. Appl. Poly. Sci. 9,* 1994 (1965); R. R. Dileone, *J. Polym. Sci. A1, 8,* 609 (1970).
22. L. I. Samarai, W. A. Bondar and G. I. Derkach, *Angew. Chem. 79,* 897 (1967).
23. U. v. Gizycki and G. Oertel, *Angew. Chem. 80,* 363 (1968).
24. R. Richter and H. Ulrich, Liebigs Ann. in press.
25. H. Ulrich, B. Tucker, F. A. Stuber and A. A. R. Sayigh, *J. Org. Chem. 33,* 3928 (1968).
26. W. Fink, Ber. *97,* 1424 (1964).

HENRI ULRICH, *The Upjohn Company; Donald S. Gilmore Research Laboratories, North Haven, Connecticut.*

CHAPTER 3

EFFECT OF ISOCYANATE STRUCTURE ON PHYSICAL PROPERTIES AND DEGRADATION OF ONE-COMPONENT URETHANE COATINGS

INTRODUCTION

The relationships between chemical structure and physical properties of polyurethane coatings have been the subject of a number of papers as well as of some recent reviews [1-6]. However, relatively little information is published on systematic investigations relating physical and environmental properties of urethane coatings to variations in isocyanate structure. The problem of yellowing in urethane coatings based on aromatic isocyanates has placed a serious limitation on these coating systems in a number of applications where light and U.V. stability are prerequisites for their acceptance. The gradual introduction of aliphatic di and polyisocyanates into coating systems possessing good light stability has greatly increased the scope of potential applications for urethane coatings [7-12]. Hence, it was considered of interest to compare the effect of aliphatic aralkyl and aromatic diisocyanates on certain mechanical, adhesive and environmental properties including heat and U.V. stability. This investigation dealt specifically with these effects in one-component, moisture-cure urethane coatings.

MATERIALS

The raw materials used in the preparation of the coatings are listed in Table 1. In addition to the isocyanates listed in Table 1, a 50/50 equivalent weight mixture of a blend of TDI and XDI was also employed. With the exception of polyols which were dried prior to usage, the materials were used as received from the suppliers without any further purification.

PROCEDURE

The polyols were dried at 80°C. at a pressure of 1.5mm for 3 hours in order to remove traces of moisture. The polyurethanes were prepared at an NCO/OH ratio of 1.8/1.

The triol (6.0 equiv.) and the respective diisocyanates (14.4 equiv., 75% solids in xylene) were allowed to react, with nitrogen sparging, at 50°C. until the free —NCO content indicated complete reaction between the hydroxyl groups of the triol and the isocyanate groups. At this stage, diol (2.0 equiv.) was added and the reaction continued until the free -NCO content indicated completion of the reaction. The required amount of xylene was then added to prepare a 50% resin solution. Films (2 mil) were drawn on glass and aluminum panels after 24 hours of standing and were allowed to cure by exposure to atmospheric moisture at room temperature. In the case of the polyurethane made from $H_{12}MDI$, stannous octoate (0.02% by wt. based on total solids) was used as a cstalyst to accelerate the urethane formation. No catalyst was needed for the preparation of the polyurethane employing XDI. The processing data for the preparation of the resin vehicles are shown in Table 2.

Under the reaction conditions employed; this formulation is expected to give an idealized polyurethane structure as shown in Figure 1.

MECHANICAL PROPERTIES

The mechanical properties of films of 2.0 mil thickness were obtained from samples prepared by using a 1/8 inch die. The stress-strain properties were determined on an Instron tester at a crosshead speed of 0.2 in./min. These results are summarized in Table 3.

These data show the films derived from aromatic isocyanates (TDI and MDI) to have substantially higher tensile strengths than the films based on cycloaliphatic and aralkyl diisocyanates $H_{12}MDI$ and XDI). The films based on 50/50 mixtures of TDI and XDI (denoted by X—TDI) are intermediate. Of particular interest is the high extensibility of films based on XDI which is also exhibited in blends with TDI.

The effect of lower crosslink density on the stress-strain properties of these films is clearly apparent, with the films chain-extended with Diol 763 exhibiting lower tensile strength and higher elongation than those chain-extended with Diol 425.

The Sward hardness of these films, as shown in Table 3 exhibited the expected decrease with decreasing crosslink density, with the XDI-based materials having the lowest values. The Gardner impact tests (see Table 3) show all of these films to possess in general very good impact strengths with the films with lower crosslink density (and greater extensibility) exhibiting higher impact strength.

Table 1. Raw Materials Used in Polyurethane Coatings.

Designation	Composition	Supplier
Triol 422[a]	Poly (oxypropylene) adduct of Trimethyiolpropane	Wyandotte Chemicals Corp.
Diol 425	Poly (oxypropylene) glycol	Wyandotte Chemicals Corp.
Diol 763	Poly (oxypropylene) glycol	Wyandotte Chemicals Corp.
XDI	Xylylene Diisocyanate 70/30 mixture of meta and para isomers	Takeda Chemical Co.
TDI	Tolylene Diisocyanate, 80/20 mixture of 2,4/2,6-isomers	Wyandotte Chemicals Corp.
MDI	4,4′ Diphenylmethane Diisocyanate	Mobay Chemical Co.
$H_{12}MDI$	4,4′ Dicyclohexylmethane Diisocyanate	Allied Chemical Corp.
T-9	Stannous Octoate	M & T Chemicals Corp.

[a]The number in the designation is the experimentally determined average mol. wt.

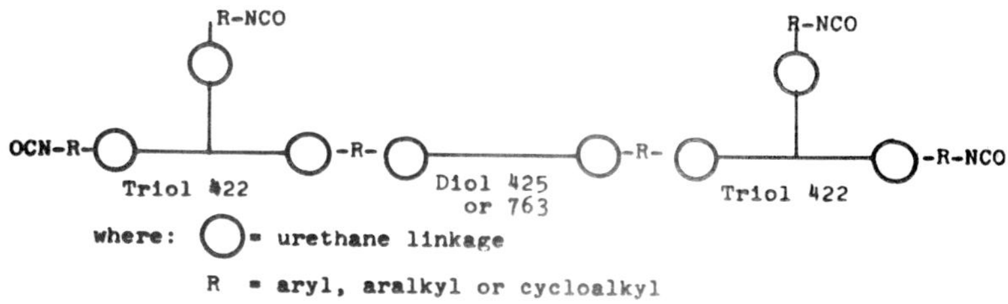

Figure 1. Idealized Chain Extended Prepolymer Structure.

ABRASION RESISTANCE OF FILMS

The cured films (2 mil thickness) were stripped from glass panels by immersing in cold water for 24 hours. The abrasion resistance of the dried films was determined on a Taber Abrader for 200 cycles using a CS-17 wheel under a load of 500 gms. These results are shown in Table 4.

Most of these films exhibited similar resistance to abrasion; the films with lower crosslink density and higher elongation showed a noticeable trend toward lower abrasion loss (with the exception of the X-TDI-based film).

BOND STRENGTH OF THE COATINGS TO COTTON FABRIC

Lap-joints were prepared using one-inch wide strips of white army-duck cloth which was preshrunk and prewashed. The ends of the strips were soaked in the polyurethane solutions, then pressed together lightly to squeeze off excess polymer solutions. These joints were cured by atmospheric moisture for one week and then conditioned at 100% relative humidity for two weeks to insure complete curing. The adhesion was tested on an Instron using a

Table 2. Processing Data For Polyurethanes.

Designation	Isocyanate group content at the end of First Stage %	Second Stage %
XDI - Triol 422 - Diol 425	16.06	10.26
XDI - Triol 422 - Diol 763	16.06	9.09
TDI - Triol 422 - Diol 425	16.83	10.66
TDI - Triol 422 - Diol 763	16.83	9.40
MDI - Triol 422 - Diol 425	13.35	8.76
MDI - Triol 422 - Diol 763	13.35	7.89
H_{12}MDI - Triol 422 - Diol 425	12.93	8.52
H_{12}MDI - Triol 422 - Diol 763	12.93	7.70
X-TDI[a] - Triol 422 - Diol 425	16.44	10.46
X-TDI[a] - Triol 422 - Diol 763	16.44	9.24

[a]X-TDI is a 50-50 mixture of XDI and TDI, based on equivalents.

Table 3. Physical Properties of Polyurethane Films.

Urethane Sample	Tensile Strength, Psi	Elongation %	Sward Hardness	Impact Strength, in.-lbs. Direct	Impact Strength, in.-lbs. Indirect
XDI - Triol 422 - Diol 425	4465	72.0	18	>80	>80
XDI - Triol 422 - Diol 763	3539	196.0	14	>80	>80
TDI - Triol 422 - Diol 425	8000	4.0	34	50	40
TDI - Triol 422 - Diol 763	5209	5.0	32	>80	>80
MDI - Triol 422 - Diol 425	7862	6.0	30	>80	>80
MDI - Triol 422 - Diol 763	6080	7.0	28	>80	>80
H_{12}MDI - Triol 422 - Diol 425	5760	8.0	38	50	30
H_{12}MDI - Triol 422 - Diol 763	5120	10.5	24	>80	>80
X-TDI - Triol 422 - Diol 425	6400	10.0	26	60	70
X-TDI - Triol 422 - Diol 763	3878	52.0	24	>80	>80

crosshead speed of 2″/min. The bond strength in terms of pounds per inch adhesion width is shown in Table 5.

These tests indicated that the coatings based on H_{12}MDI showed the best performance, followed by the coatings based on MDI and TDI. The XDI-based coatings exhibited the lowest bond strength. However, coatings based on blends of TDI and XDI exhibited somewhat higher bond strength than would have been expected from a 50/50 blend of XDI and TDI.

Table 4. Taber Abrasion Tests.

Urethane Sample	Wt. Loss In Mgms. Per 200 Cycles*
XDI - Triol 422 - Diol 425	4.1
XDI - Triol 422 - Diol 763	2.7
TDI - Triol 422 - Diol 425	6.0
TDI - Triol 422 - Diol 763	3.5
MDI - Triol 422 - Diol 425	4.0
MDI - Triol 422 - Diol 763	3.3
H_{12}MDI - Triol 422 - Diol 425	10.0
H_{12}MDI - Triol 422 - Diol 763	4.5
X-TDI - Triol 422 - Diol 425	3.6
X-TDI - Triol 422 - Diol 763	3.6

*CS-17 wheel and 500 gms load.

Table 5. Bond Strengths of Cotton Fabric - Urethanes.

Urethane Sample	Bond Strength-lbs/in.
XDI - Triol 422 - Diol 425	5.1
XDI - Triol 422 - Diol 763	3.0
TDI - Triol 422 - Diol 425	12.6
TDI - Triol 422 - Diol 763	9.0
MDI - Triol 422 - Diol 425	13.75
MDI - Triol 422 - Diol 763	12.5
H_{12}MDI - Triol 422 - Diol 425	15.5
H_{12}MDI - Triol 422 - Diol 763	14.0
X-TDI - Triol 422 - Diol 425	10.8
X-TDI - Triol 422 - Diol 763	8.25

RESISTANCE TO SOLVENTS

The resistance to solvents such as dichloromethane and chloroform was determined from the swelling ratio of the films at -10°C. The films were immersed in the solvent for 6 hours then allowed to dry for 2 minutes at room temperature and then weighed. No appreciable change in the swelling ratio was observed on examining the films after 24 hours of immersion.

The average molecular weight per crosslink, $\overline{Mc}$ (as an indication of crosslink density) was determined by the method of Sandridge et al [13], using a modified expression as follows:

$$\overline{Mc} = \frac{F}{\overline{F} - 2} (\overline{E}_p + \overline{E}_I)$$

where $\overline{F}$ is the average functionality of the reactants (polyols and isocyanates), $\overline{E}_p$ is the average equivalent weight of polyols, and $\overline{E}_I$ is the average equivalent weight of diisocyanate. No corrections were made for the presence unsaturation in the polyols or any possible side reactions of the isocyanate groups. The $\overline{Mc}$ and the swelling ratios in chloroform and dichloromethane at -10°C. for 6 hours are shown in Table 6.

No distinct trend seemed to be apparent from the swelling data. In general, the films showed a somewhat higher swelling in chloroform as compared to methylene chloride. However, with the exception of the H_{12}MDI based films, which exhibited a higher swelling ratio than the other films, relatively little difference was observed.

RESISTANCE TO YELLOWING–FADE-OMETER TEST

In order to compare the relative stability of the coatings, unstabilized as well as stabilized, these films were subjected to continuous exposure in a Fade-ometer at 52°C. and 32% relative humidity for 200 hours. The yellowing index on the films was obtained before and after exposure using a colorimeter (Model LS Color-Eye, Instrument Development Laboratories, Attleboro, Mass.).

The various stabilizers used are shown in Table 7. The increase in the yellowing index after exposure is listed in Table 8. Previous work by Schollenberger et al, Beachell et al, and others [14-18] have shown that yellowing involves an oxidative process promoted by U.V. light, the mechanism of which depends on the nature of the diisocyanate.

It can be readily observed that the films based on the aliphatic isocyanates exhibited superior performance over the aromatics. In the group of aliphatics, the H_{12}MDI showed less yellowing than XDI, even when both were stabilized. Among the aromatics, TDI-based films were superior to the MDI-based films, presumably due to the lack of the methylene group in the TDI.

The effect of various antioxidant-UV absorbers is also shown in Table 8. Their effects were very pronounced with the aromatic isocyanate-based films, while it was only very slight in the case of XDI-based films and did not appear to have any beneficial effect on the H_{12}MDI-based films. Of particular interest were the data from films based on blends of TDI and XDI. Even the unstabilized films exhibited a surprisingly low degree of yellowing which was even further reduced by the use of antioxidant-UV absorber pairs; these values were very close to those from films based solely on XDI.

Additional evidence for the beneficial effect of the antioxidant-UV absorber pairs in the stabilization of films based on aromatic isocyanate and XDI is shown by the changes in the tensile strength values, by calculating the differences between the values before and after exposure to 200 hours of Fade-ometer testing as seen in Table 9.

Table 6. Mc And Swelling Ratios In $CHCl_3$ And CH_2Cl_2 At -10°C. For 24 Hrs.

Urethane Sample	Mc (Theoretical)[a]	Swelling Ratio, Vs/Vo, %	
		$CHCl_3$	CH_2Cl_2
XDI - Triol 422 - Diol 425	626	124	116
XDI - Triol 422 - Diol 763	862	124	115
TDI - Triol 422 - Diol 425	616	121	110
TDI - Triol 422 - Diol 763	853	120	112
MDI - Triol 422 - Diol 425	669	121	115
MDI - Triol 422 - Diol 763	906	128	112
H_{12}MDI - Triol 422 - Diol 425	678	133	127
H_{12}MDI - Triol 422 - Diol 763	914	142	129
X-TDI - Triol 422 - Diol 425	621	125	114
X-TDI - Triol 422 - Diol 763	858	122	116

[a]Using a modified expression of Sandridge et al, J. Chem. Eng. Data, *5*, 495 (1960).

Table 7. Stabilizers Used In Polyurethane Coatings.

Antioxidants	Composition	Supplier
Irganox 1076	n-Octadecyl 3-(3′,5′-di-t-butyl-4′-hydroxylphenyl) propionate	Geigy Ind. Chem.
Ionol	2,6-Di-t-butyl-p-cresol	Shell Chemical
M-24	Mixed hindered phenols	Pitt-Consol. Chem.
U.V. Absorber		
Tinuvin 327	2-(3′,5′-ditert-butyl-2′-hydroxyphenyl)-5-chlorobenzotriazole	Geigy Ind. Chem.

THERMAL DECOMPOSITION OF POLYURETHANE FILMS

Studies of the thermal decomposition of substituted urethanes have shown that three general types of reactions take place:

1. Dissociation to isocyanate and alcohol

$$RNHCOOR' \rightleftharpoons RNCO + R'OH$$

2. Formation of primary amine and olefin

$$RNHCOOCH_2CH_2R' \longrightarrow RNH_2 + CO_2 + R'CH{=}CH_2$$

3. Formation of secondary amine

$$RNHCOOR' \longrightarrow RNHR' + CO_2$$

Substituted urea groups in the polyurethanes are known to dissociate to the corresponding isocyanate and amine at elevated temperatures:

$$RNHCONHR' \rightleftharpoons RNCO + R'NH_2$$

In this study, the investigation of the thermal decomposition of the polyurethane films was carried out by means of thermal gravimetric analysis (TGA) and differential thermal analysis (DTA) using a du Pont Model 900 Differential Thermal Analyzer and Model 950 Thermogravimetric Analyzer. The samples were tested at a heating rate of 20°C/min. using nitrogen and air at a flow rate of 0.5 LPM.

The thermograms are shown in Figures 2 to 6.

Films based on Diol 425 and XDI exhibited an exotherm starting at about 180°C, while the corresponding film based on Diol 763 resulted in an exotherm at 160°C, indicating that oxidation was faster with higher concentration of ether groups.

Table 8. Fade-Ometer Tests.

		Increase in Yellowness Index		
Urethane Sample	**Unstabilized**	**1% Irganox 1076 + 1% Tinuvin 327**	**1% Ionol + 1% Tinuvin 327**	**1% M-24 + 1% Tinuvin 327**
XDI - Triol 422 - Diol 425	3	3	2	2
XDI - Triol 422 - Diol 763	4	3	3	3
TDI - Triol 422 - Diol 425	24	12	8	8
TDI - Triol 422 - Diol 763	26	10	9	9
MDI - Triol 422 - Diol 425	28	22	18	26
MDI - Triol 422 - Diol 763	45	14	21	18
H_{12}MDI - Triol 422 - Diol 425	1	2	2	1
H_{12}MDI - Triol 422 - Diol 763	1	1	1	1
X-TDI - Triol 422 - Diol 425	8	4	4	4
X-TDI - Triol 422 - Diol 763	6	3	3	3

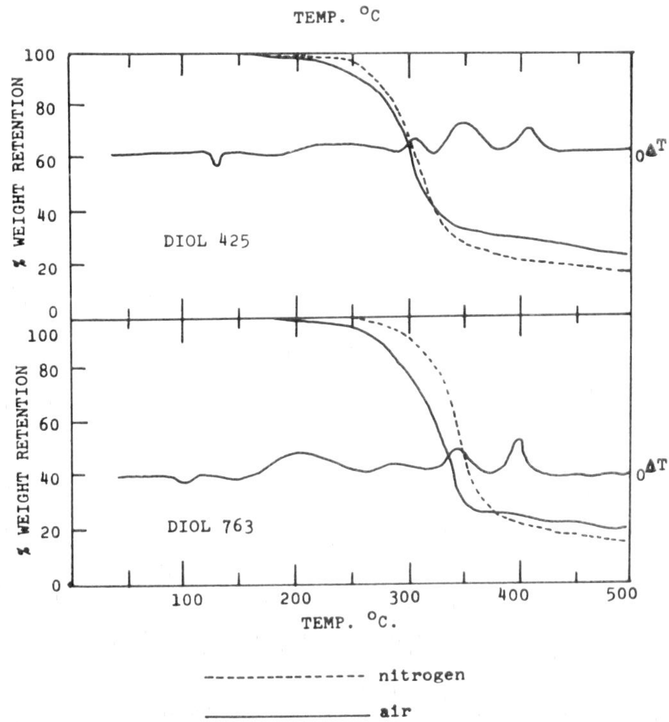

Figure 2. DTA-TGA From XDI-Triol 422-Diol Systems.

Table 9. Fade-Ometer Tests—Loss in Tensile Strength[a]

Urethane Sample	Decrease In Tensile Strength; % Unstabilized	1% Irganox 1076 + 1% Tinuvin 327	1% Ionol + 1% Tinuvin 327	1% M-24 + 1% Tinuvin 327
XDI - Triol 422 - Diol 425	7	2	4	6
XDI - Triol 422 - Diol 763	22	10	9	5
TDI - Triol 422 - Diol 425	53	35	44	42
TDI - Triol 422 - Diol 763	52	11	11	14
MDI - Triol 422 - Diol 425	29	21	25	28
MDI - Triol 422 - Diol 763	44	26	23	25
H_{12}MDI - Triol 422 - Diol 425	6	5	6	2
H_{12}MDI - Triol 422 - Diol 763	4	1	2	0
X-TDI - Triol 422 - Diol 425	26	15	10	14
X-TDI - Triol 422 - Diol 763	8	4	2	2

[a]Crosshead speed of 0.2"/min.

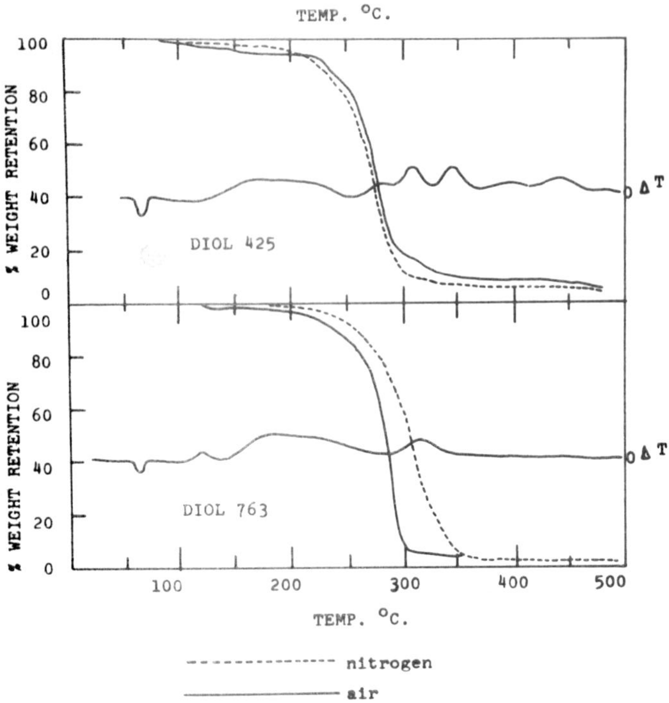

Figure 3. DTA-TGA From TDI-Triol 422-Diol Systems.

The urethane from XDI and Diol 425 showed a loss in weight in air and nitrogen starting at 165°C. This was close to the exotherm shown in the DTA thermogram. The urethane from Diol 763 showed a loss in weight in air starting at about 200°C. In nitrogen atmosphere, this loss in weight started at about 250°C.

With TDI and Diol 425 an exotherm occurred at 130°C, indicating that urethanes from TDI were less resistant to oxidation than those from XDI. Again, the urethane from Diol 763 had an exotherm at a lower temperature (115°C).

The urethane from TDI and Diol 425 exhibited a weight loss at 75°C. which was probably due to loss of solvent or moisture. The major weight losses at 165°C. and 215°C. were probably the result of oxidation. The urethane from Diol 763 exhibited major weight losses at 200°C. in air, probably due to oxidation. The weight loss that started at 225°C. in nitrogen was caused by decomposition.

With MDI and Diol 425 an endotherm at about 100° was probably caused by the loss of solvent or moisture. An exotherm occurred at about 260°C, indicating a high resistance to oxidation. The urethane from Diol 763 again had an exotherm at a lower temperature (200°C).

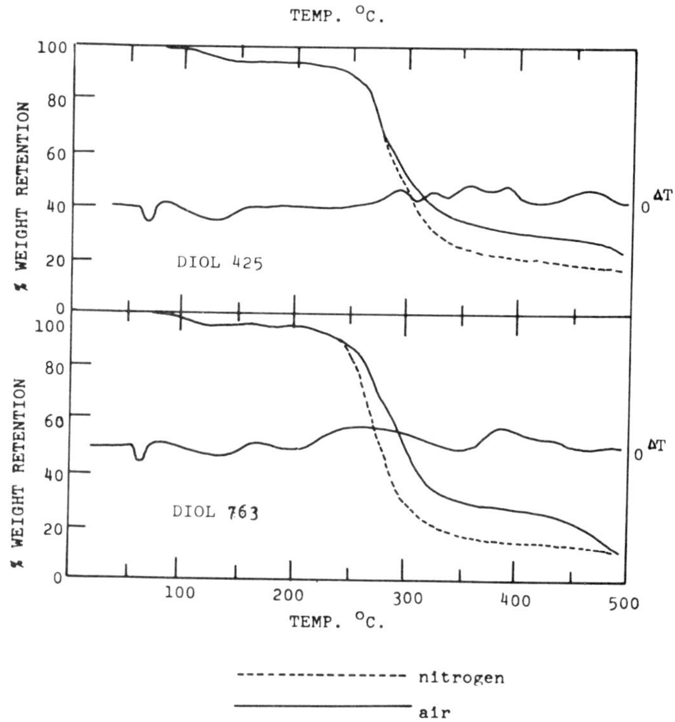

Figure 4. DTA-TGA From MDI-Triol 422-Diol Systems.

After initial weight losses due to evaporation of solvent or moisture, the urethane from MDI and Diol 425 showed a major weight loss due to oxidation at 250°C, which was fairly close to the major weight loss temperature for the urethane from the Diol 763 (225°C.). It is interesting to note that the weight losses in nitrogen atmosphere were significantly greater than the losses in air.

With H_{12}MDI and Diol 425, after a small exotherm at 130°C, a major exotherm began at 200°C; with the Diol 763 the major exotherm started at somewhat below 200°C. Thus H_{12}MDI was somewhat more resistant than the urethane from the other aliphatic diisocyanate, XDI.

The major weight loss for the urethane from H_{12}MDI and Diol 425 started at about 175°C. and was close to the temperature of weight loss for the XDI urethane. The urethane from the Diol 763 appeared to suffer a major weight loss at a similar temperature.

The urethane from 50-50 XDI-TDI and Diol 425 started to oxidize at about the same temperature as the urethane from TDI (130°C.). The polymer from Diol 763 exhibited a surprising increase in exotherm temperature to 160°C. (much better than TDI and equivalent to XDI alone).

The major weight loss for the urethane from 50-50 XDI-TDI and Diol 425

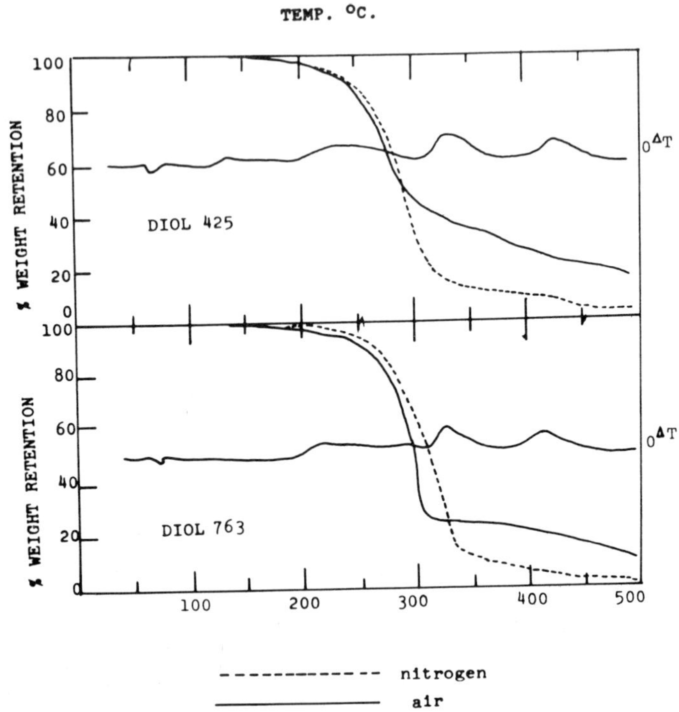

Figure 5. DTA-TGA From H_{12}MDI-Triol 422-Diol Systems.

started at 200-225°C., somewhat higher than the temperature of weight loss from the XDI or TDI urethane. A significantly lower temperature of major weight loss resulted from the Diol 763 urethane (160°C.).

THERMAL DECOMPOSITION OF URETHANES AS MEASURED BY CO_2 EVOLUTION AT 300°C.

In inert atmosphere such as nitrogen, urethanes will decompose at high temperatures to form CO_2 and secondary amines:

$$\mathrm{RNH\overset{O}{\overset{\|}{C}}OR'} \xrightarrow{\Delta} CO_2 + \mathrm{RNHR'}$$

The thermal degradation of a number of urethanes was compared by pyrolyzing the samples at 300°C. and sweeping out the products by means of a stream of nitrogen free from H_2O, oxygen and CO_2. The gaseous products were passed through a dry ice-acetone trap to condense and eliminate all but the CO_2, which was passed into an Ascarite tube. The tube was weighed periodically and the rate of degradation was determined. The rate constants for the CO_2 evolution at 300°C. were calculated using the initial urethane concentration derived from the experimentally determined total amount of

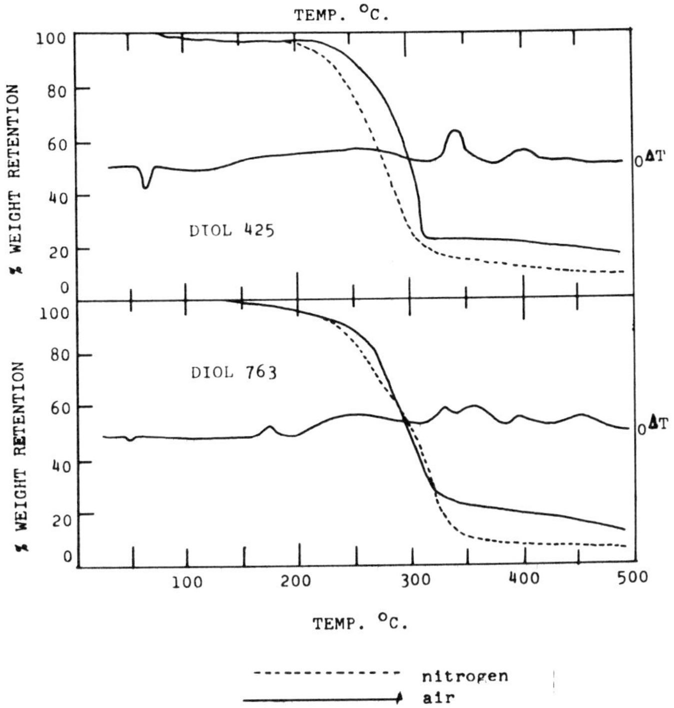

Figure 6. DTA-TGA From X-TDI-Triol 422-Diol Systems.

generated CO_2.

The rate of CO_2 evolution in an inert atmosphere (N_2) follows first order kinetics as shown in Table 10 and Figures 7 and 8.

It can be observed that the aromatic diisocyanate-based films exhibited the highest rate of degradation, with the TDI-based films showing a higher rate than the corresponding MDI-based films.

The H_{12}MDI-based films had the lowest rate of decomposition with the XDI-based films exhibiting the second lowest rate. It was interesting to note that the rate of degradation of films based on blends of XDI and TDI was nearly as low as that of films based on XDI alone.

Hence, the conclusions which can be drawn from the data based on CO_2 evolution are as follows:

1. Urethanes from H_{12}MDI show the greatest resistance to decomposition in an inert atmosphere with those from XDI having only a slightly lower resistance.
2. Urethanes from TDI under the same conditions exhibit the highest rate of decomposition.
3. Urethanes from MDI show a rate of decomposition intermediate between XDI and TDI.

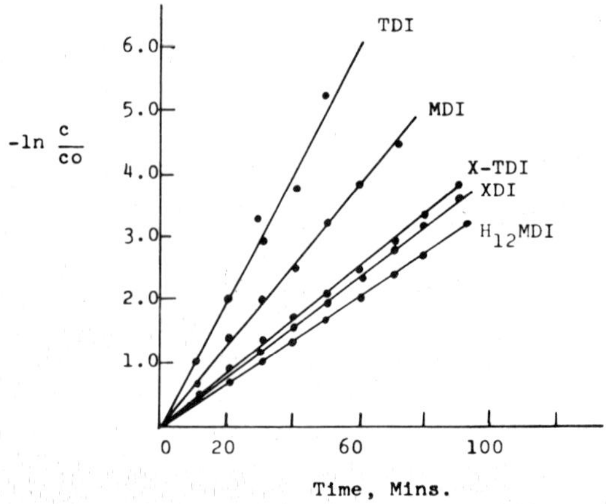

Figure 7. First Order Reaction Plots for CO_2 Evolution at 300°C. In Nitrogen for Diisocyanate-Triol 422-Diol 425 Systems.

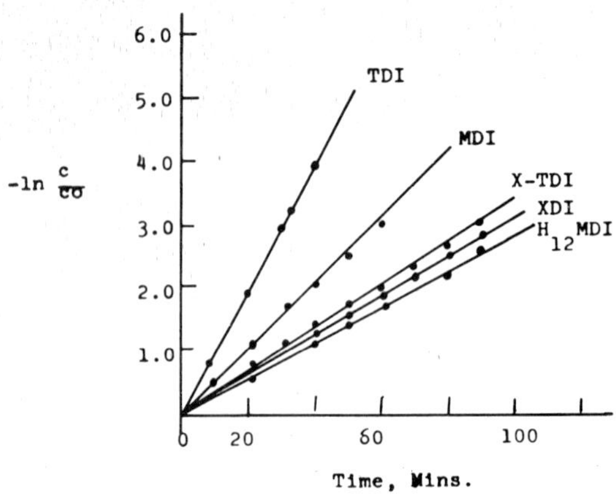

Figure 8. First Order Reaction Plots for CO_2 Evolution at 300°C. In Nitrogen Atmosphere. Diisocyanate Triol 422-Diol 763 Systems.

4. Urethanes made from 50/50 combination of XDI and TDI exhibit a rate of decomposition which approximates the data from XDI alone which is substantially lower than the arithmetic mean of both.

REFERENCES

1. A. J. Damusis, J. M. McClellan, and K. C. Frisch, Off. Digest *32,* 251 (1960).
2. J. H. Saunders and K. C. Frisch, "Polyurethanes" Part I Chemistry, Interscience Wiley, New York (1962), pp. 323,326.

3. J. H. Saunders and K. C. Frisch, "Polyurethanes" Part II Technology, Interscience Wiley, New York (1967), Chapter X.
4. M. E. Bailey, G. C. Toone, G. S. Wooster, and E. G. Bobalek, *Off. Digest 32,* 984 (1960).
5. A. Lowe, Chapter 7 in "Advances in Polyurethane Technology," edited by J. M. Buist and H. Gudgeon, Interscience-Wiley, New York (1968).
6. A. Damusis, and K. C. Frisch, Chapter 12 in "Film-Forming Compositions," Vol. I, Part I, edited by R. R. Myers and J. S. Long, Marcel Dekker, New York (1967).
7. H. Gruber, *J. Oil Colour Chemists' Assoc., 48,* 1069 (1965).
8. T. Kanzawa and K. Naito, *Japan Chem. Quart. 3,* 38 (1967).
9. J. M. Cross, S. H. Metzger, Jr., and C. D. Campbell, U. S. Pat. 3,351,650 (to Mobay Chem. Co.), Nov. 7, 1967.
10. R. L. White and G. A. Hudson, Chapter 13 in "Polyurethane Technology," edited by P. F. Bruins, Interscience Wiley, New York, 1969.
11. General Mills, Bulletin CDS-3-65, July 23, 1965.
12. K. A. Pigott, E. R. Wells and G. A. Hudson, "Light-Stable Polyurethanes," paper presented at the Organic Coatings Technology Polymer Conference at Wayne State University, June 1967.
13. R. L. Sandridge, A. S. Morecroft, E. E. Hardy, and J. H. Saunders, *J. Chem. and Eng. Data, 5,* 495 (1960).
14. C. S. Schollenberger and K. Dinbergs, *SPE Trans. 1,* 31 (1961).
15. C. S. Schollenberger, H. Scott and G. R. Moore, *Rubber World, 137,* 549 (1958).
16. H. C. Beachell and C. P. Ngoc Son, *J. Appl. Polymer Sci., 7,* 2217 (1963).
17. H. C. Beachell and C. P. Ngoc Son, *J. Appl. Polymer Sci., 8,* 1089 (1964).
18. L. V. Nevskii, O. G. Tarakanov and V. K. Belyakov, *Soviet Plastics* (Engl. Transl.), No. *7,* 45 (1966).

Table 10. Thermal Degradation: - First Order Rate Constants For Evolution of CO_2 At 300°C. In Nitrogen.

Urethane Sample	$K \times 10^{-2}$, $min.^{-1}$
XDI - Triol 422 - Diol 425	3.88
XDI - Triol 422 - Diol 763	3.50
TDI - Triol 422 - Diol 425	9.75
TDI - Triol 422 - Diol 763	9.67
MDI - Triol 422 - Diol 425	6.33
MDI - Triol 422 - Diol 763	5.25
H_{12}MDI - Triol 422 - Diol 425	3.25
H_{12}MDI - Triol 422 - Diol 763	2.83
X-TDI - Triol 422 - Diol 425	4.17
X-TDI - Triol 422 - Diol 763	3.38

K. C. FRISCH, S. L. REEGEN and L. P. RUMAO, *Polymer Institute, University of Detroit, Detroit, Michigan 48221*

CHAPTER 4

THERMOPLASTIC POLYURETHANE HYDROLYSIS STABILITY

INTRODUCTION

Thermoplastic polyurethane elastomers are the newest, fastest growing members of the very versatile polyurethane family. These polymers are essentially linear in structure and can be processed by the methods customarily used with thermoplastics, including solution applications. Yet the polymers exhibit many of the properties of a rubbery vulcanizate including excellent strength, toughness, and abrasion resistance, as well as high extensibility and elasticity. This unique physical state is due to strong interchain attractive forces such as hydrogen bonding, van der Waals forces, and chain entanglements. For these reasons the polymers have been referred to as "virtually crosslinked" elastomers [1].

The natural and artificial weathering of a representative, thermoplastic, aromatic, poly(ester-urethane) elastomer in temperate climate has been studied and reported [2]. From these studies the dominant mechanism of polymer degradation was concluded to be an ultraviolet-initiated autoxidation process.

Although there is some published information on the hydrolytic stability of thermoplastic polyurethane elastomers [7] the amount is meager. The present paper attempts to help fill this void. It will also supplement hydrolysis stability information published on vulcanized cast and millable gum polyurethane elastomers [3, 4, 5, 6, 7, 8, 9].

1. THERMOPLASTIC POLYURETHANE ELASTOMERS

(A) Monomers

Most of the thermoplastic polyurethane elastomers on the market today are prepared from three types of monomers. These are, of course, used to make other polyurethane types also.
They are: a diisocyanate, a macroglycol, and a glycol chain extender (Figure 1).

In the formulas of Figure 1 the nature of the groups R, R′, and R″ is quite important to the properties, both physical and chemical, of thermoplastic polyurethanes made from the corresponding monomers.

Figure 2 shows the nature of the R groups in two diisocyanates, MDI and TDI, which are now used in making thermoplastic polyurethane elastomers

(1)	Diisocyanate	OCN—R—NCO
(2)	Macroglycol	HO—R′—OH
(3)	Glycol Chain Extender	HO—R″—OH

Figure 1. Monomer Types Used to Make Thermoplastic Polyurethane Elastomers.

R in MDI **(diphenylmethane–4,4'–diisocyanate)**

R in TDI **(2,4–isomer) (2,6–isomer)** **(tolylene diisocyanate)**

Figure 2. R Structures in the Diisocyanate Component, OCN-R-NCO, Commonly Used to Make Thermoplastic Polyurethanes.

$$\left[CH_2CH_2CH_2CH_2-\underset{\sim\sim}{O-CO}-CH_2CH_2CH_2CH_2-\underset{\sim\sim}{CO-O} \right]_{5\text{-}10} CH_2CH_2CH_2CH_2-$$

R' in poly(tetramethylene adipate) glycol, a polyester glycol (MW 1000-2000)

$$-\underset{\sim}{O}\left[CH_2CH_2CH_2CH_2-\underset{\sim}{O} \right]_{14\text{-}28} CH_2CH_2CH_2CH_2-\underset{\sim}{O}-$$

R' in poly(tetramethylene oxide) glycol, a polyether glycol (MW 1000-2000)

Figure 3. R' Structures in the Macroglycol Component, HO-R'-OH, Commonly Used to Make Thermoplastic Polyurethane Elastomers.

commercially. Note that R is an aromatic structure in both of these diisocyanates. The outstanding strength, toughness, and abrasion resistance of most solid polyurethane elastomers now on the market includes a strong contribution from such aromatic rings in the polymer chains.

Some aliphatic and cycloaliphatic diisocyanates are gaining in importance as polyurethane monomers since they yield more color-stable polymers.

Now let's look at what we call the "macroglycol" building blocks used in making thermoplastic polyurethane elastomers.

Figure 3 shows the R 'structures in two representative macroglycols, a polyester glycol and a polyether glycol. The upper formula shows R 'to be the tetramethylene adipate radical as it occurs in poly(tetramethylene adipate) glycol, a polyester glycol. The lower formula shows R 'to be the oxytetramethylene radical as it occurs in poly(oxytetramethylene) glycol, a polyether glycol.

As is shown, these structural units are connected end to end and repeated several times in each macroglycol molecule. Consequently, the macroglycols are very large molecules as monomers go, having molecular weights of about 1000-2000. But they are still small enough to melt to fluid liquids at less than 50°C. I have underlined the ester and ether groups in the formulas of Figure 3. I will have more to say about these groups later.

Another type of useful, commercially available polyester glycol is

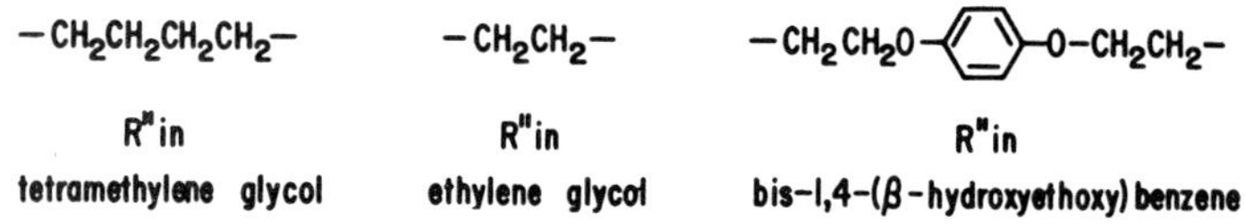

Figure 4. R″ Structure in the Glycol Chain Extender Component, HO-R″ -OH, Commonly Used to Make Thermoplastic Polyurethane Elastomers.

poly(ε-caprolactone) glycol. We will later see the chemical equations involved in the formation of all of these macroglycol types.

Macroglycol structure choices are deliberate on the part of the polymer chemist. For the selected chain structures are flexible. And since the macroglycol component comprises 60-80% of the derived polyurethane weight the polyurethane elastomer inherits this characteristic and is also flexible and elastomeric.

Figure 4 shows the structure of the R″ groups in three types of chain extenders commonly used to make thermoplastic polyurethane elastomers. The formula on the left represents the tetramethylene radical as it occurs in tetramethylene glycol, which is sometimes called 1,4-butanediol. The formula in the middle represents the ethylene radical as it occurs in ethylene glycol. And the formula on the right represents the bis-1,4-butanediol. The formula in the middle represents the ethylene radical as it occurs in ethylene glycol. And the formula on the right represents the bis-1,4-(oxy-ethylene) benzene radical as it occurs in bis-1,4-(β-hydroxyethoxy) benzene.

The structure of the R″ radical is also important in determining thermoplastic polyurethane elastomer properties. Its greatest influence is on polymer physical properties. By varying the type and amount of such chain extenders in thermoplastic polyurethane recipes one can regulate polyurethane hardness, modulus, and elasticity. For example, the use of tetramethylene glycol or ethylene glycol, both open chain glycols, as polyurethane chain extender produces softer, more extensible polymers than are obtained with equivalent amounts of the aromatic-structured glycol, bis-1,4-(β-hydroxyethoxy) benzene.

(B) Polymer Structure

Now, having reviewed the monomer types which are commonly used to make thermoplastic polyurethane elastomers, let's look at the preparation and structure of these polymers.

An equation representing the formation of a thermoplastic polyurethane elastomer is shown in Figure 5. In practice a diisocyanate may be added to a blend of a macroglycol and a small glycol chain extender and the mixture allowed to polymerize. We call these random melt polymerization conditions.

(C) Polymer Properties

The polymer product represents a unique physical state. For it has many of the properties of a strong rubbery vulcanizate but it is not really vulcanized.

$$\underset{\text{(diisocyanate)}}{2\ \mathrm{OCN{-}R{-}NCO}} + \underset{\text{(macroglycol)}}{\mathrm{HO{-}R'{-}OH}} + \underset{\text{(small glycol)}}{\mathrm{HO{-}R''{-}OH}} \longrightarrow$$

$$-\underline{\mathrm{O}}\!\left(\underline{\mathrm{CONH}}\mathrm{{-}R{-}}\underline{\mathrm{NHCO{-}O}}\mathrm{{-}R'{-}}\underline{\mathrm{OCONH}}\mathrm{{-}R{-}NHCO{-}O{-}R''{-}O}\right)_n$$

Figure 5. Thermoplastic Polyurethane Elastomers.

Strong interchain attractive forces such as hydrogen bonding, van der Waals forces, and chain entanglements are believed to account for this "virtually crosslinked" state. These attractive forces can be overcome by heating the polymer sufficiently, but they reappear when it cools. This, of course, allows the polymer to be processed by the methods normally used with thermoplastics such as milling, calendering, extrusion, compression-, transfer-, and injection-molding, etc. Excellent physical properties are thus inherent in the polymer as received from the supplier. There is no need for mixing in curing agents or of any vulcanization step whatever.

But the polymers can be compounded. They accept pigment loadings well and make useful blends with a variety of other polymers.

In addition to these characteristics the thermoplastic polyurethanes are fully soluble in some practical organic solvents. That is to say, their interchain attractive forces can also be overcome by solvation. This allows many useful solution applications, such as in adhesives, binders, and transfer coating, whose importance cannot be overemphasized. For when the solvent is removed the original high level properties of the dissolved polymer are restored.

On the other hand the thermoplastic polyurethane elastomers show a practical degree of resistance to many common solvents, and excellent resistance to hydrocarbon oils and fuels.

The full solubility of the thermoplastic polyurethane elastomers in some organic solvents, their thermoplastic nature, and their method of preparation support the view that they are essentially linear polymers.

It is not the intent of this paper to provide detailed information on the properties of the various thermoplastic polyurethane elastomers on the market today. Some property information will appear in this paper to the extent that it is needed to illustrate the results of our hydrolysis stability studies. However, in all cases it was obtained by testing laboratory-prepared polymer samples.

II. HYDROLYSIS STABILITY STUDIES

(A) Suitability of Thermoplastic Polyurethane Elastomers for Hydrolysis Stability Studies

Of all the polyurethane elastomer types the thermoplastic polyurethane elastomers are probably the best suited for scientific environmental resistance studies. This is because the use-forms of these polymers are linear structures.

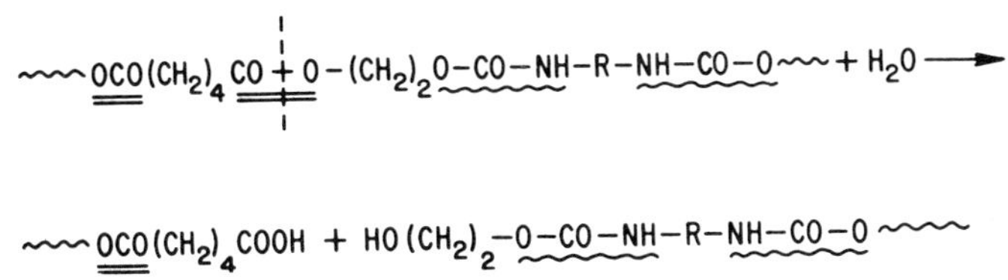

Figure 6. Poly(Ester-Urethane) Hydrolysis.

So the chemical and physical consequences of changes in the polymer chains are more easily demonstrated than in the case of covalently crosslinked polyurethanes. For example, in the case of thermoplastic poly(ester-urethane) elastomers hydrolytic degradation can be followed by noting changes in polymer viscosity and carboxylic acidity.

Moreover, the purity of the thermoplastic polyurethane elastomers can be superior since their use does not require catalysts or curing chemicals. Such agents or their residues are usually permanent and may adversely affect polymer environmental stability.

(B) Hydrolysis Stability Test

Hydrolysis is the chemical decomposition of substance by water.

Figure 6 shows water reacting with a carboxylic ester link (doubly underlined) in a poly(ester-urethane) chain. The reaction breaks the polymer chain at the point of attack producing two shorter chains. One of these chains as you see now ends in a hydroxyl group (—OH). The other ends in a carboxyl group (—CO_2H) which is acidic. This acidic carboxyl group speeds up the further hydrolysis of the polyester segments in the poly(ester-urethane) and the degradation becomes autocatalytic [3, 4, 6]

Our accelerated hydrolysis stability test is intended to speed up the same chemical reaction—carboxylic ester link hydrolysis—that dominates in poly(ester-urethane) hydrolytic degradation in normal service. But there is another hydrolytic degradation reaction possible.

The urethane linkage which I have singly underlined in the formulas of Figure 6 is also properly referred to as a carbamate linkage. That is, it is an ester of (a substituted) carbamic acid. This urethane or carbamate linkage can also be hydrolyzed, but less readily than the carboxylic ester linkages.

Hydrolysis of urethane linkages in polyurethane chains produces two shorter new chains, one ending in a hydroxyl group (—OH) and the other in an amino group (—NH_2). But unless a poly(ester-urethane) elastomer contains special sterically hindered carboxylic ester linkages in its chains the dominant reaction in its hydrolysis in normal service is cleavage of the carboxylic ester groups as shown in Figure 6.

The relative merits of polyurethane hydrolysis stability tests conducted under humid vs. total immersion conditions have been discussed elsewhere.

Polyurethane Composition	II	III
Polyurethane Exposure (Weeks/70°C H_2O)	1	1
Polyurethane Water Content, Exposed Sample (phr)	1.92	1.46

II 2.00 MDI/Poly(tetramethylene adipate) glycol (MW 953, AN 3.84)/1,4-BDO
III 2.00 MDI/Poly(hexamethylene adipate) glycol (MW 986, AN 2.42)/1,4-BDO

Figure 7. Water Absorption by Thermoplastic Polyurethane Samples.

One view [4, 6] proposes that poly(ester-urethane) hydrolysis conditions which remove acidic polymer degradation fragments are less severe than those which do not, since residual acidic fragments autocatalyze the hydrolysis. Thus, poly(ester-urethane) foams are reported to resist hot running water better than high humidity at the same temperature. And thin films of poly(ester-urethane) elastomer better than thick films (presumably in the hot running water test and not the hot high humidity test).

In a later study 70°C, high humidity aging was the favored accelerated hydrolytic exposure condition because of its greater similarity to service conditions [7]. But the same study showed little difference in the degradation index (tensile retention) of unstabilized, peroxide-cured, poly(ester-urethane) elastomer whether exposed to high humidity at 70°C or totally immersed in water at the same temperature. Degradation rates in humid air and under water were found to be the same in still another study [9].

In another study both humid aging and total immersion were used [6].

In a more recent study some critical aspects of high temperature-high humidity aging such as heat transfer to the sample and water condensation on the sample are emphasized [8].

The hydrolytic exposure condition used in the present study involved total immersion of the test samples in static 70°C water.

The exposure was primarily intended to allow comparison of the relative stabilities of various thermoplastic polyurethane elastomer structures towards hydrolysis. The degree of polymer chemical and physical changes due to the exposure was used to judge this stability.

The use of an elevated test temperature not only accelerated the test but also was considered insurance against the complication of polymer degradation by microbial action. This problem was encountered at or near room temperature in a previous study [7].

Our exposures were performed under water to minimize the possibility of any contribution to polymer degradation by thermally accelerated polymer autoxidation.

Autoxidation of poly(ether-urethanes) in humid air hydrolysis tests was suspected in an earlier study [9]. And our earlier work showed that autoxidation of thermoplastic poly(ester-urethanes) can occur at elevated temperatures [2].

In practice the hydrolytic exposure of the various thermoplastic poly-

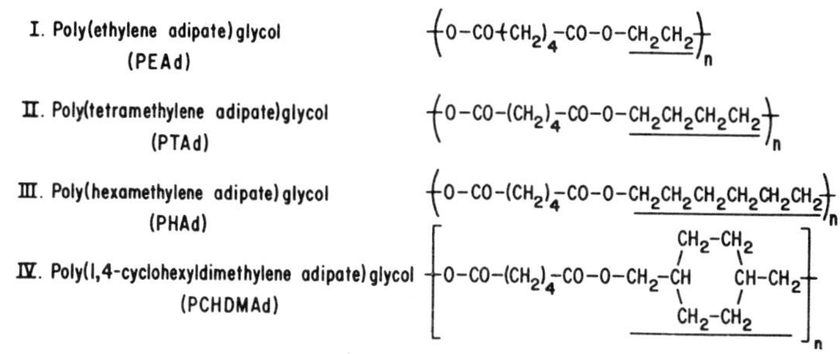

Figure 8. Macroglycol Repeat Unit Structures.

urethane elastomer compositions involved immersion of 0.025" thick micro-dumbbell samples in glass tubes freshly filled with distilled water for each exposure. The tubes were stoppered and placed in a 70° ± 0.5°C multicavity aluminum block for 1, 2, and 3 or more weeks. Samples were exposed in triplicate.

The samples were then removed and desiccated over Drierite for 24 hours. This was done to assure removal of absorbed immersion water in excess of the equilibrium level that would normally be absorbed in subsequent sample conditioning. The conditioning treatment was then applied to both exposed and unexposed samples before test. This procedure was necessary since absorbed water exerts a plasticizing effect in polyurethanes which is reflected in stress-strain properties [6]. So to accurately assess the effects of hydrolytic exposure by noting changes in stress-strain properties, efforts were made to have similar, fairly normal moisture levels in exposed and unexposed samples at the time of test.

The macroglycol component structures of Figure 7 are shown in Figure 8. The data of Figure 7 show that composition III with its higher methylene group concentration absorbed less water than composition II after one week immersion in 70°C water.

After the exposure the absorbed water was removed from the test pieces by 24 hour desiccation over Drierite. The samples were then reconditioned at 50% R. H. and 77°F for 24 hours with the aim of getting them to about the same, fairly normal, absorbed moisture levels.

(C) Hydrolysis Stability of Thermoplastic Poly(Ester-Urethanes)
(1) Effect of Polyester Macroglycol Structure.

Several different poly(ester-urethane) compositions were prepared, tested, and their hydrolysis resistance compared. To allow valid comparisons the polymer compositions in this series were all made from two moles of diisocyanate (MDI), one mole of macroglycol (variable), and one mole of chain extender (1,4-butanediol). The same random, melt polymerization conditions were used for all polymer preparations.

Figure 8 shows the structural variations of the macroglycol components we

Polyurethane	Macroglycol Name	Mol. Wt.	Acid No.
(A)	Poly(Ethylene Adipate)Glycol	980	0.75
(B)	Poly(Ethylene Adipate)Glycol	1017	0.08
(C)	Poly(Tetramethylene Adipate)Glycol	989	3.66
(D)	Poly(Tetramethylene Adipate)Glycol	1046	2.50
(E)	Poly(Tetramethylene Adipate)Glycol	1110	0.60
(F)	Poly(Tetramethylene Adipate)Glycol	1098	0.00
(G)	Poly(Tetramethylene Adipate)Glycol	1001	0.10
(H)	Poly(Hexamethylene Adipate)Glycol	986	2.42
(I)	Poly(Hexamethylene Adipate)Glycol	1148	0.61
(J)	Poly(1,4-Cyclohexyldimethylene Adipate)Glycol	1190	0.60

Figure 9. Macroglycol Component of Various Poly(Ester-Urethane) Compositions.

compared in this portion of our hydrolysis stability study. Note that as one descends the table the composition of the repeating unit in the macroglycol chains changes. Each repeating unit is seen to contain two more methylene ($-CH_2-$) units than the one above. I have underlined these varying methylene units in Figure 8.

It will be seen that macroglycols of about the same molecular weights were used in this series. So as one descends Figure 8 the concentration of methylene groups, relative to other groups in the polymer chains such as ester and urethane groups, increases. With this change the derived polyurethanes would be expected to be increasingly hydrophobic and thus more hydrolysis-resistant in this sequence.

Figure 7 data indicate that the higher methylene group concentration of a polyurethane based on poly(hexamethylene adipate) glycol causes it to absorb less water than a polyurethane based on poly(tetramethylene adipate) glycol during a week immersion in 70°C water.

It should also be noted that although the methylene group concentrations of macroglycol structures I, II, and III were varied by using different open chain aliphatic glycols to make the corresponding polyester glycols, a spacious, more rigid, cycloaliphatic glycol was used to make macroglycol structure IV. It will be seen that this macroglycol, poly(1,4-cyclohexyldimethylene adipate) glycol, produced a considerably harder polyurethane than the other macroglycols. The increased hardness would be expected to contribute to reduced polyurethane permeability, and to further enhance polyurethane hydrolysis resistance.

Figure 9 lists further information on the polyester macroglycol components used to make the thermoplastic polyurethanes of this series.

Macroglycol average molecular weights were determined by end group analysis. Note that the molecular weights are similar for the series at 1085 ± 105 units.

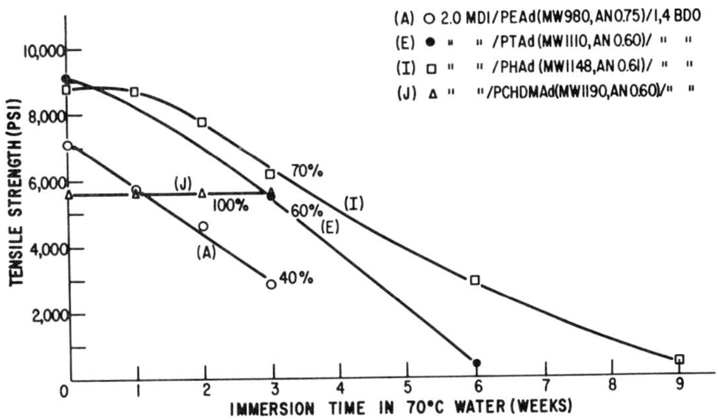

Figure 10. Hydrolysis Resistance of Various Thermoplastic Poly-(Ester-Urethane) Compositions Made From Moderate Acid Number Macroglycols.

Acid number is the mg. of potassium hydroxide required to neutralize the carboxylic acidity of 1.0g. of the polyester glycols. Take careful note of the acid numbers of the various macroglycol samples. For we find that this acid number is a dominant factor in the hydrolytic stability of derived thermoplastic polyurethane elastomers [10].

And now let's examine some examples of the effect of polyester macroglycol structure on thermoplastic poly(ester-urethane) hydrolysis stability.

Figure 10 compares the hydrolysis stability of several different compositions made from macroglycols having about the same moderate acid numbers. It shows that polyurethane (1) based on poly(hexamethylene adipate) glycol, and having higher methylene group concentration, has better hydrolytic stability than polyurethane (A) which is based on poly(ethylene adipate) glycol.

The percentage figure associated with each curve in the plots is the portion of the polyurethane's original tensile which is retained after immersion for three weeks in 70°C water. It is seen that polyurethane (1) retained 70% of its original strength and polyurethane (A) only 40% in this exposure.

Polyurethane (E), based on poly(tetramethylene adipate) glycol, with a methylene group concentration intermediate to that of polyurethanes (A) and (I), also showed intermediate hydrolysis stability. It retained 60% of its original tensile strength on three week immersion in 70°C water.

The outstanding performance of polyurethane (J), which is based on poly-(1,4-cyclohexyldimethylene adipate) glycol, shows the effect of its higher methylene group concentration as well as its greater hardness.

Figure 11 provides more information on the properties of the polymers in Figure 10 and the exposure effects. It is seen that the dilute solution viscosities of the different polymers were reduced about the same degree as their tensile strengths by the exposure.

Macroglycol Name	PEAd	PTAd	PHAd	PCHDMAd
Macroglycol Mol. Wt.	980	1100	1148	1190
Macroglycol Acid No.	0.75	0.60	0.61	0.60
Polyurethane	(A)	(E)	(I)	(J)
Polyurethane DSV	1.3505 (44)*	1.233	1.160 (73)	0.697 (90)
Polyurethane T_2 (°C)	154	157	153	142
Polyurethane Hardness (Shore)	88A	85A	89A	60D
Polyurethane Tensile (psi)	7100 (40)	9100 (60)	8800 (70)	5600 (100)
Polyurethane Elongation (%)	650 (119)	485 (147)	515 (123)	355 (107)
Polyurethane 300% Modulus (psi)	1100 (100)	1200 (83)	1000 (110)	4800 (79)

* Number is original value: number in parentheses is % of original value retained after three week immersion in 70°C water.

Figure 11. Effect of Macroglycol Structure on Hydrolytic Stability of Thermoplastic Poly(Ester-Urethanes) Made From Moderate Acid Number Macroglycols.

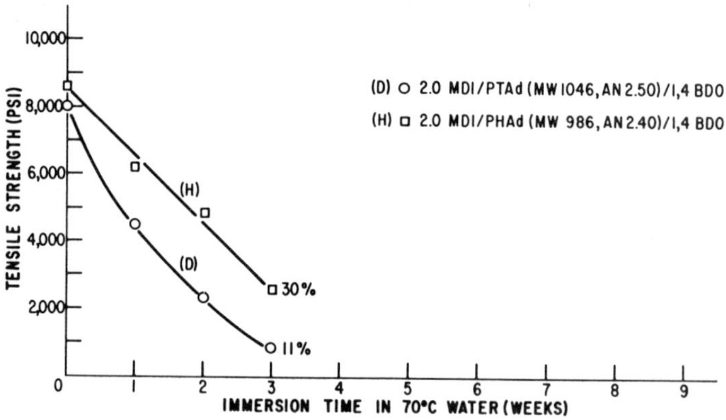

Figure 12. Hydrolysis Resistance of Various Thermoplastic Poly-(Ester-Urethane) Compositions Made From Higher Acid Number Macroglycols.

Macroglycol Name	PTAd	PHAd
Macroglycol Mol. Wt.	1046	986
Macroglycol Acid No.	2.50	2.42
Polyurethane	(D)	(H)
Polyurethane DSV	1.008 (36)*	1.058 (42)
Polyurethane T_2 (°C)	154	147
Polyurethane Hardness (Shore)	90A	82A
Polyurethane Tensile (psi)	8000 (11)	8600 (30)
Polyurethane Elongation (%)	550 (119)	560 (131)
Polyurethane 300% Modulus (psi)	1200 (58)	1200 (75)

* Number is original value: number in parentheses is % of original value retained after three week immersion in 70°C water.

Figure 13. Effect of Macroglycol Structure on Hydrolysis Stability of Thermoplastic Poly-(Ester-Urethanes) Made From Higher Acid Number Macroglycols.

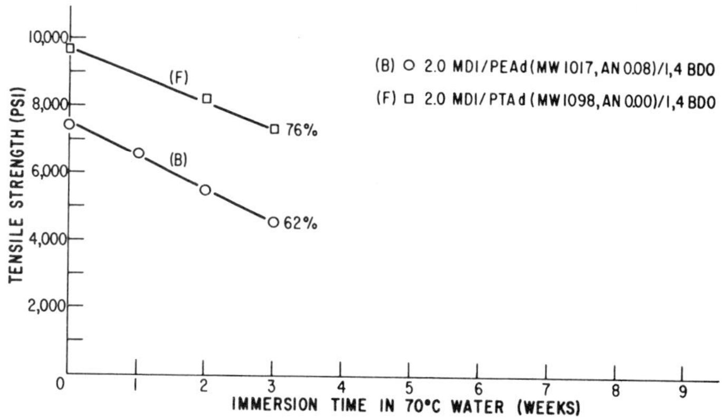

Figure 14. Hydrolysis Resistance of Various Thermoplastic Poly(Ester-Urethane) Compositions Made Low Acid Number Macroglycols.

The T_2 temperature of Figure 11 and subsequent figures is the melt flow temperature value obtained from a rheology test [1]. It represents the temperature at which the polymer flows well through an orifice under specific conditions.

DSV stands for the dilute solution viscosity of the polymer determined at a polymer concentration of 0.40 g per 100 ml. of dimethyl formamide solvent at 25°C. All polymers were gel-free.

Figure 12 compares the effect of methylene group concentration in macroglycols of similar, higher acid number on the hydrolysis resistance of derived polyurethanes. Again it is seen that polyurethane (H), based on poly(hexamethylene adipate) glycol, shows better hydrolytic stability than polyurethane (D) which is based on poly(tetramethylene adipate) glycol, whose methylene group concentration is lower. Polyurethane (H) retained 30% of its original tensile strength and polyurethane (D) 11% after three week, 70°C water immersion.

Figure 13 provides more information on the properties of the polymers of Figure 12 and the exposure effects. Comparison with counterpart polyurethanes (E) and (I) of Figures 10 and 11 will show that those polymers, made from lower acid number polyester glycols, were more stable to hydrolysis. But the relative order of stability was the same.

Figure 14 compares the effect of methylene group concentration in macroglycols of similar low acid number on the hydrolysis resistance of derived polyurethanes. Once more it is seen that polyurethane (F) which is based on poly(tetramethylene adipate) glycol, and has the greater methylene group concentration, is more hydrolysis-resistant than polyurethane (B) which is based on poly(ethylene adipate) glycol. Polyurethane (F) retained 76% of its original tensile strength and polyurethane (B) 62% after three week, 70°C, water immersion.

These examples of the effect of methylene group concentration on poly

Macroglycol	Name	PEAd	PTAd
Macroglycol	Mol. Wt.	1017	1098
Macroglycol	Acid No.	0.08	0.00
Polyurethane		(B)	(F)
Polyurethane	DSV	0.824 (74)*	1.501 (75)
Polyurethane	T_2 (°C)	136	179
Polyurethane	Hardness (Shore)	86A	92A
Polyurethane	Tensile (psi)	7400 (62)	9700 (76)
Polyurethane	Elongation (%)	655 (128)	490 (115)
Polyurethane	300% Modulus (psi)	900 (111)	1300 (84)

* Number is original value: number in parentheses is % of original value retained after three week immersion in 70°C water.

Figure 15. Effect of Macroglycol Structure on Hydrolysis Stability of Thermoplastic Poly-(Ester-Urethanes) Made From Low Acid Number Macroglycols.

$$\text{excess } HO(CH_2)_2OH + HO_2C(CH_2)_4CO_2H \longrightarrow$$

(ethylene glycol) (adipic acid)

$$HO\left[(CH_2)_2OCO(CH_2)_4COO\right]_n(CH_2)_2OH + HO\left[(CH_2)_2OCO(CH_2)_4COO\right]_nH + H_2O$$

(zero acid number) (measureable acid number)

(poly(ethylene adipate)glycol)

Figure 16. Polyester Glycol Formation.

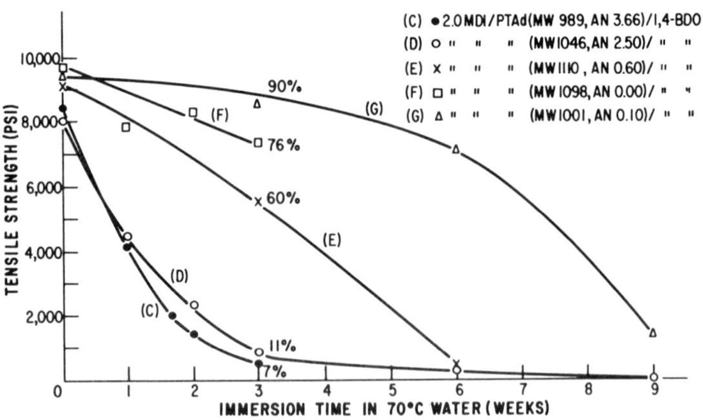

Figure 17. Hydrolysis Resistance of Thermoplastic Poly(Ester-Urethanes) Made From Poly(Tetramethylene Adipate) Glycol of Different Acid Numbers.

(ester-urethane) hydrolytic stability are less convincing than other examples in this paper due to the greater hardness, stiffness, and (most likely) molecular weight of the superior polymer, (F). But they are consistent with the other patterns and are likely valid.

Figure 15 provides further information on the properties of the polymers of Figure 14 and the exposure effects. As in the cases of polyurethanes (A), (E), (H), (I), (J), and (G) which is yet to be discussed, the loss in the dilute solution viscosities of (B) and (F) parallels their loss in tensile strength. Comparison with counterpart polyurethanes (A) and (E) of Figures 10 and 11 will show that those polymers, made from moderate acid number polyester glycols were less stable to hydrolysis than polyurethanes (B) and (F).

(2) Effect of Polyester Macroglycol Acidity.

Although the examples presented and discussed in this paper show that the composition of the polyester macroglycol influences the hydrolytic stability of derived thermoplastic poly(ester-urethane) elastomers, we find that it is not the dominant factor. Rather it is the degree of acidity of the polyester macroglycol used to make the poly(ester-urethane). This may have been apparent to you in some of the examples already discussed. The purpose of the present section is to make comparisons which clearly show it to be true.

First, why should a polyester glycol be acidic? It is due to an incomplete esterification reaction in its preparation. A common way of making polyester macroglycols is to condense a dibasic acid, such as adipic acid, with excess of a glycol, such as ethylene glycol, applying heat to speed the reaction and vacuum to remove the water generated. This reaction is shown in Figure 16.

Most of the polyester macroglycol molecules formed terminate at both ends in hydroxyl (—OH) groups and have zero acid number as intended. But frequently a few of the polyester macroglycol molecules formed terminate at one end of the chain in free, unesterified carboxyl (—CO_2H) groups. These carboxyl groups are acidic and their level represents the acid number of the polyester macroglycol. Let's see just how much this acid number affects the hydrolysis stability of derived thermoplastic poly(ester-urethanes).

Figure 17 compares the hydrolysis stability of a series of thermoplastic poly(ester-urethane) elastomers. The compositions of the polymers of this series are the same. All were made from two moles of the same diisocyanate, MDI, one mole of the same glycol chain extender, 1,4-butanediol, and one mole of the same polyester macroglycol, poly(tetramethylene adipate) glycol. But the acid number of the macroglycol was varied. And notice what an effect this had on the hydrolytic stability of the derived thermoplastic poly(ester-urethanes).

Polyurethanes (C) and (D), made from the higher acid number polyester glycols retained only 7% and 11%, respectively, of their original tensile strengths on three week immersion in 70°C water. Polyurethane (E) made from intermediate acid number polyester macroglycol retained much more

Macroglycol Name	Poly(Tetramethylene Adipate)Glycol				
Macroglycol Mol. Wt.	989	1046	1110	1098	1001
Macroglycol Acid No.	3.66	2.50	0.60	0.00	0.10
Polyurethane	(C)	(D)	(E)	(F)	(G)
Polyurethane DSV	0.904 (31)*	1.032 (30)	1.233	1.501 (75)	1.136 (88)
Polyurethane T_2 (°C)	160 —	154 —	157	179 —	161 —
Polyurethane Hardness (Shore)	88A —	90A —	85A	92A —	87A —
Polyurethane Tensile (psi)	8400 (7)	8000 (11)	9100 (60)	9700 (76)	9400 (90)
Polyurethane Elongation (%)	550 (35)	550 (119)	485 (147)	490 (115)	475 (119)
Polyurethane 300% Modulus (psi)	1300 (68)	1200 (58)	1200 (83)	1300 (84)	1800 (100)
Polyurethane Mn**	—	—	47,500	—	—

* Number is original value: number in parentheses is % of original value retained on three weeks immersion in 70°C water.
** Number average molecular weight, osmometry, DMF, 37°C.

Figure 18. Effect of Macroglycol Acid Number on Hydrolysis Stability of Thermoplastic Poly(Ester-Urethane) Elastomers - 2.0 MDI/PTAd/1,4-BDO.

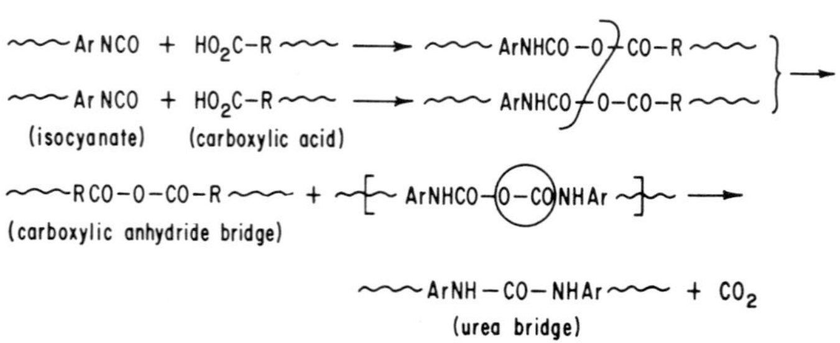

Figure 19. Reaction of Aromatic Isocyanate Group with Carboxyl Group.

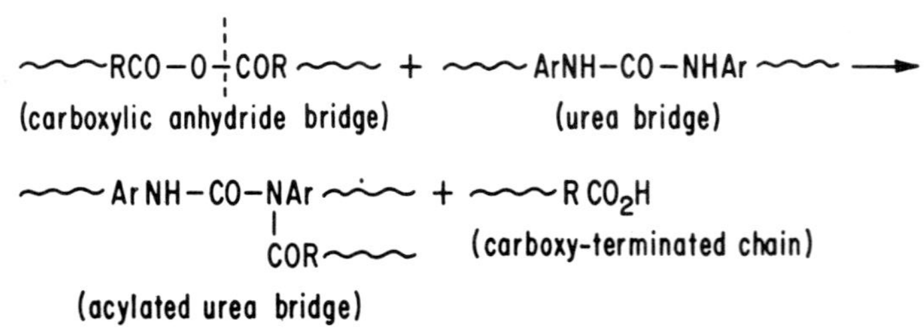

Figure 20. Possible Reaction of Anhydride Bridge with Urea Bridge.

(60%) of its original tensile strength. And polyurethanes (F) and (G), made from low acid number polyester macroglycol retained high percentages (76%, 90%) of their original strengths in the exposure.

Figure 18 provides further information on the properties of the polymers of Figure 17 and the exposure effects.

The data of Figures 17 and 18 clearly show that in the case of poly (tetramethylene adipate) glycol the macroglycol acidity adversely affects the hydrolysis stability of derived thermoplastic poly(ester-urethanes) in proportion to the degree of acidity. The same pattern would be seen if one were to compare the performances of certain other polyurethane compositions in this test. For example, polyurethanes based on poly(ethylene adipate) glycol, such as polyurethane (A) of Figures 10 and 11 vs. polyurethane (B) of Figures 14 and 15. And polyurethanes based on poly(hexamethylene adipate) glycol, such as polyurethane (H) of Figures 12 and 13 vs. polyurethane (I) of Figures 10 and 11.

(3) Explanation of Polyester Macroglycol Acidity Effect.

The effect of parent macroglycol composition on the hydrolysis stability of derived thermoplastic poly(ester-urethanes) seemed clear and rational. But the role of polyester macroglycol acid number in the matter was not apparent at first.

We know that the carboxyl group contains an active hydrogen atom which reacts readily with the isocyanate group [11]. In the case of the carboxyl and aromatic isocyanate groups the reaction could proceed as in Figure 19 [12]. If this reaction does occur in thermoplastic poly(ester-urethane) formation then the polymer chains would contain a few urea bridges and carboxylic anhydride bridges. The anhydride bridges should register as carboxylic acidity when the acid number determination is applied to the thermoplastic poly(ester-urethane), which we have done.

But do the polyurethane chains contain anhydride bridges? The carboxylic anhydride linkage is reported to react with the urea linkage at temperatures above about 160°C [12]. This temperature is commonly achieved in polyurethane melt polymerizations. One would expect the N-acylation reaction shown in Figure 20 to occur in the thermoplastic poly(ester-urethane) if there were carboxylic anhydride and urea bridges present in the polymer chains. Moreover, acid anhydride and urea may also react to give an amide and carbon dioxide [17]:

$$\text{ArNH—CO—NHAr} + \text{R'—CO—O—CO—R'} \rightarrow 2\ \text{ArNHCO—R'} + \text{CO}_2$$

These reactions would substantially reduce the carboxylic acidity of the polyurethane if they occurred. But we have found that the acidity of the thermoplastic poly(ester-urethane) approximates the level expected if none of the free, unesterified, terminal carboxyl groups of the parent polyester

$$AN_U = \frac{MW_E \times AN_E}{MW_{URU}}$$

Where AN_U = Acid no. of polyurethane
AN_E = Acid no. of polyester glycol
MW_E = Molec. Wt. of polyester glycol
MW_{URU} = Molec. Wt. of polyurethane repeat unit

Figure 21. Equation Relating Macroglycol and Derived Polyurethane Acidities.

	Acid Number		
	Polyester Macroglycol	Poly(Ester-Urethane)	
Polymer Preparation	Found	Theory	Found
1	2.50	1.59	1.52
2	3.29	2.05	1.95
(D)	2.50	1.60	1.51
4	2.50	1.60	1.50
5	0.10	0.07	0.12
6	0.10	0.07	0.18
(G)	0.10	0.06	0.27
(F)	0.00	0.00	0.16
9	0.00	0.00	0.15

Figure 22. Relation of Polyester Macroglycol Acid Number to Thermoplastic Poly(Ester-Urethane) Acid Number.

macroglycol had reacted during the thermoplastic poly(ester-urethane) formation. So we suspect that the anhydride and urea bridge forming reactions of Figure 19 do not take place.

Figure 21 shows a relationship that enables one to determine the theoretical acid number of a thermoplastic poly(ester-urethane), assuming that none of the carboxyl groups of the parent polyester macroglycol are neutralized by chemical reaction during polyurethane formation. Use of this relationship requires that one know the acid number of the parent polyester macroglycol, its molecular weight, and the molecular weight of the polyurethane repeat unit. We feel that we can measure the acid number with an accuracy of 0.1 acid number unit.

Figure 22 relates the acid number of the parent polyester macroglycol to the acid number found in several derived thermoplastic poly(ester-urethane) elastomers. These polymers were all prepared under random melt polymerization conditions from two moles of MDI and one mole each of poly(tetramethylene adipate) glycol and 1,4-butanediol.

It is seen that the acid numbers found for the polyurethanes agree fairly well—within about 0.1 unit—with the theoretical values, which assume no chemical neutralization of carboxyl groups during polyurethane formation.

In the higher acid number macroglycol cases the polyurethane acid number values found are about 95% of the theoretical values. In the low acid number macroglycol cases the polyurethane acid number values found exceed the

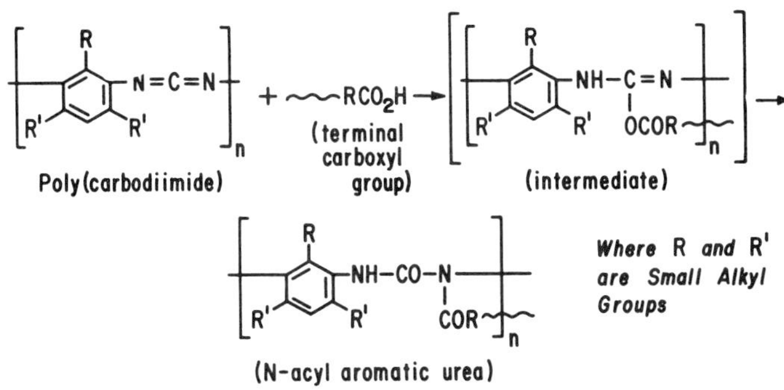

Figure 23. Reaction of a Poly(Carbodiimide) with Carboxyl Groups.

theoretical values by 0.1 to 0.2 units. And this equals or exceeds the accuracy of the acid number determination. Nevertheless, we do know that whatever its precise level, in these cases the acid number is quite low.

Although there are small, absolute differences between the theoretical and found acid numbers, the facts suggest that the terminal carboxyl groups of a parent polyester macroglycol do not participate in the polyurethane polymerization reaction to any extent but largely survive this reaction intact.

The reasons for this must include their scarcity, relative to the hydroxyl groups, in the polymerization recipe. Their inability to successfully compete for isocyanate in a stoichiometrically balanced isocyanate-active hydrogen system with the 17 fold more reactive alcoholic hydroxyl groups [13] as well as other more reactive trace impurities such as water must also be a factor.

Reference to Figure 22 shows considerable differences in the acid numbers of polyurethanes (D) and (G) or (F). Figures 12, 13, 17 and 18 for (D); 17 and 18 for (G); and 14, 15, 17 and 18 for (F) are convincing evidence of the dependence of thermoplastic poly(ester-urethane) elastomer hydrolysis resistance on the acid number of the polymer. The preparation of poly(ester-urethanes) of improved hydrolysis stability from low acid number polyester macroglycols has been patented [10].

(4) Stabilization and Destabilization by Additives.

The autocatalytic nature of the poly(ester-urethane) hydrolysis reaction (Figure 6) was discussed earlier in this paper. Some mechanism for neutralizing the carboxyl-containing fragments of the hydrolyzing polyester segments as they are generated should retard the hydrolysis and stabilize the polymer.

A very effective class of materials for this purpose has been described [4, 5]. They are the poly(carbodiimides).

Figure 23 shows one type of sterically hindered, aromatic poly(carbodiimide) structure which is claimed to be a hydrolysis stabilizer for poly(ester-

Polyurethane		(D)	(D) + 2.0 phr PCD[a]	(G)	(G) + 2.0 phr PCD	(K)	(K) + 2.0 phr PCD	(F)	(F) + 1.0 phr Stearic Acid
DSV	(Original)	1.032	1.085	1.136	1.108	0.763	0.938	1.501	1.178
DSV									
Retained	(3 weeks)[b]	30%	88%	88%	95%	48%	93%	75%	30%
Retained	(6 weeks)	16%	87%	68%	92%	24%	89%	—	—
Retained	(9 weeks)	11%	79%	36%	85%	16%	88%	—	—
Acid No.	(Original)	1.51	—	0.24	—	1.60	0.25	0.16	—
Acid No.	(3 weeks)	—	—	0.54	—	5.40	0.22	—	—
Acid No.	(6 weeks)	—	—	1.11	—	15.0	0.17	—	—
Acid No.	(9 weeks)	—	—	3.80	—	29.2	0.16	—	—

(a) PCD is a poly(carbodiimide) supplied by Naftone, Inc., N. Y.
(b) Immersion time in 70°C water.

Figure 24. Changes in Thermoplastic Poly(Ester-Urethane) Viscosity and Acid Number on Hydrolysis as Affected by Added Poly(Carbodiimide) and Carboxylic Acid. 2.0 MDI/PTAd/1,4-BDO.

urethanes) [4, 5]. Attributes are reported to include: non-volatility; resistance to extraction; fairly good physiological inertness; low $-N=C=N-$ self polymerization tendencies; and good compatibility with plastics and elastomers containing ester groups [4].

Figure 23 also shows the suspected mechanism of the carboxyl-carbodiimide reaction. The carboxyl group when formed adds across one C = N double bond of the carbodiimide group to form a transitory intermediate. Rearrangement to the stable, neutral, N-acyl aromatic urea structure follows. Thus, the carboxyl group has been chemically neutralized and bound into the poly(carbodiimide) structure. The same fate awaits the next carboxyl group formed by hydrolysis, etc.

It can be seen that this sequence of reactions prevents autocatalysis of the hydrolysis reaction by neutralizing the carboxyl groups as they are formed. Moreover, since each stabilizer molecule contains several carbodiimide groups it can rejoin the broken polymer chains which terminate in carboxyl groups, and thus mend many of these broken chains.

In the first data column of Figure 24 the impressive viscosity stabilization of polyurethane (D) by two parts of PCD is seen. In the second data column the greater hydrolysis stability of low acid number polyurethane (G) is apparent. Stabilization of this polymer with PCD was more effective. Note the steady increase in the acid number of unstabilized polymer (G) with exposure time.

Polyurethane (K) in the third data column was prepared from two moles of MDI, one mole of poly(tetramethylene adipate) glycol of about 1000 molecular weight and 2.5 acid number, and one mole of 1,4-butanediol, like the other polyurethanes of this study. But its solution viscosity is lower. The high initial acidity of this unstabilized poly(ester-urethane) resulted in a large increase in its acid number during the hydrolysis exposure.

As can be seen the addition of PCD to polyurethane (K) increased its solution viscosity while lowering its acid number significantly. These concurrent changes are believed to demonstrate the polymer chain mending action mentioned earlier, that is, the joining of carboxy-terminated poly(ester-urethane) chains by reaction with the stabilizer.

It is of considerable interest that the acid number of stabilized polyurethane (K) decreased, rather than increased with exposure time showing the sustained chemical function of the stabilizer in long exposures.

In the fourth data column we see the pronounced acceleration of the hydrolysis of polymer (F) by 1.0 phr of stearic acid ($C_{17}H_{35}CO_2H$) which was added to the polyurethane on the rubber mill. This acid is a long chain, water-insoluble, carboxylic acid commonly added to rubbers and plastics as a lubricant. But our work clearly shows that carboxylic acids such as stearic acid should not be added to thermoplastic poly(ester-urethanes) as lubricants, or for other purposes.

Now let's see how poly(carbodiimide) affects thermoplastic poly(ester-urethane) stress-strain property retention in the hydrolysis test.

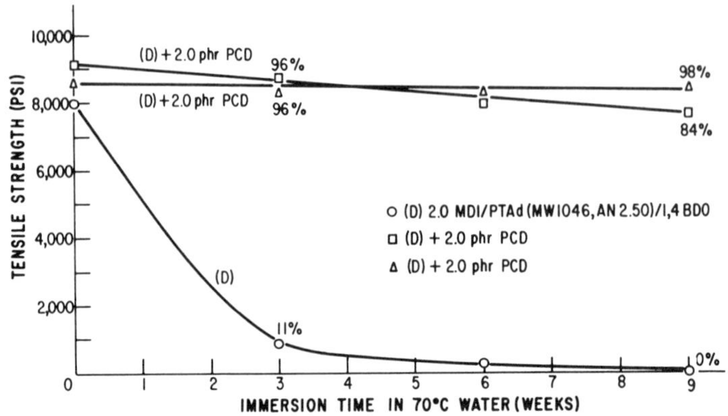

Figure 25. Effect of Added Poly(Carbodiimide) on Hydrolysis Stability of a Thermoplastic Poly(Ester-Urethane) Made From Higher Acid Number Macroglycol.

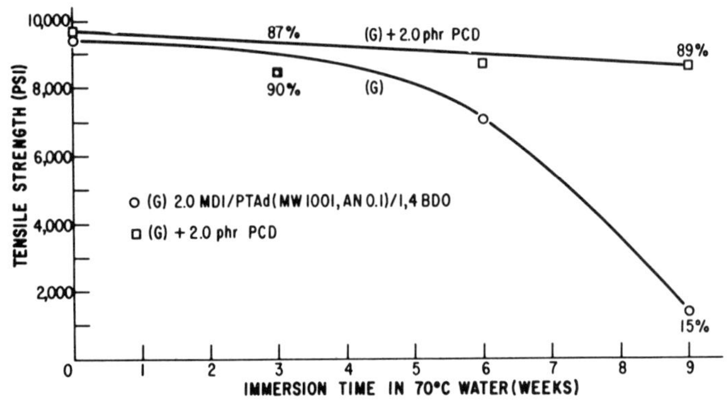

Figure 26. Effect of Added Poly(Carbodiimide) on Hydrolysis Stability of a Thermoplastic Poly(Ester-Urethane) Made From Low Acid Number Macroglycol.

Figure 25 shows the pronounced improvement in hydrolysis stability that 2.0 phr of the stabilizer PCD imparts to polyurethane (D) which was made from a higher acid number polyester marcroglycol. The PCD was added to the polyurethane on a rubber mill. Two trials gave similar results. The stabilized polymer retained 96% of its original tensile strength on three week immersion in 70°C water. The unstabilized control polymer retained only 11% in the same exposure. After nine week exposure the stabilized polymer still retained 90% of its original tensile strength while the unstabilized control polymer had disintegrated.

Figure 26 shows similar stabilizing action of 2.0 phr of the poly(carbodiimide) on polyurethane (G) which was made from low acid number polyester

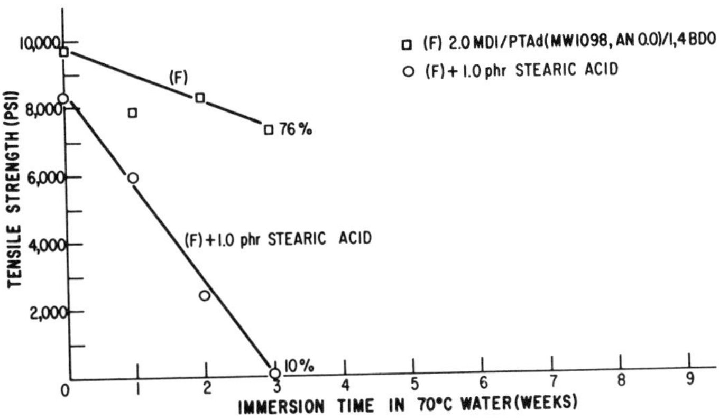

Figure 27. Effect of Added Carboxylic Acid on Hydrolysis Stability of a Thermoplastic Poly(Ester-Urethane) Made From Low Acid Number Macroglycol.

macroglycol. In this case the beneficial action of the stabilizer was not strongly apparent until about the sixth week of exposure due to the greater inherent hydrolysis stability of polyurethane (G).

After three week exposures the stabilized and unstabilized polymers showed about 90% retention of their original tensile strengths. And after nine weeks of exposure the value was still 89% for the stabilized polymer but only 15% for the unstabilized control. So the stabilizer clearly benefited the poly (ester-urethane) made from low acid number polyester macroglycol in the long exposure.

Figure 27 shows the effect of 1.0 phr of stearic acid on the stress-strain properties of polyurethane (F) in the hydrolysis stability test. Three week immersion in 70°C water reduced the tensile strength of the compound to 10% of its original value. But the stearic acid-free control polymer retained 76% of its tensile strength in the same exposure. Lower levels of stearic acid had a similar effect.

Figure 28 shows further effects of added poly(carbodiimide) stabilizer and added stearic acid destabilizer on the hydrolysis stability of the thermoplastic poly(ester-urethane) elastomers and their compounds considered in Figures 25, 26 and 27.

High acid number polyurethanes (D) and (K) disintegrated on nine week immersion in 70°C water, whereas low acid number polyurethane (G) retained its full elongation and half of its modulus after the same exposure. The benefit of poly(carbodiimide) hydrolysis stabilizer is particularly evident in the longer (six and nine week) exposures.

An increase in 300% modulus is apparent in all three polyurethane-poly (carbodiimide) compounds between the three and six week exposures. This could be another manifestation of the chain mending action of the poly(carbodiimide) stabilizer. For the stabilizer is capable of crosslinking the system

Polyurethane	(D)	(D) + 2.0 phr PCD	(G)	(G) + 2.0 phr PCD	(K)	(K) + 2.0 phr PCD	(F)	(F) + 1.0 phr Stearic Acid
T_2 (°C)	154	—	161	—	—	—	179	—
Hardness (Shore)	90A	—	87A	—	91A	90A	92A	82A
Elongation (Original) (%)	550	500	500	490	565	550	490	550
% of Original								
Elongation Retained (3 weeks)[a]	127%	122%	118%	117%	117%	114%	115%	91%
Elongation Retained (6 weeks)	7%	115%	133%	123%	13%	115%	—	—
Elongation Retained (9 weeks)	disintegrated	128%	116%	123%	2%	107%	—	—
300% Modulus (Original) (psi)	1200	2000	1900	2000	1400	1300	1300	1300
% of Original 300%								
Modulus Retained (3 weeks)	58%	65%	69%	70%	50%	100%	85%	46%
Modulus Retained (6 weeks)	—	75%	79%	80%	<29%	115%	—	—
Modulus Retained (9 weeks)	—	65%	47%	70%	<14%	115%	—	—

(a) Immersion time in 70°C water.

Figure 28. Changes in Thermoplastic Poly(Ester-Urethane) Stress-Strain Properties on Hydrolvsis as Affected by Added Poly(Carbodiimide) and Carboxylic Acid. 2.0 MDI/PTAd/1,4-BDO.

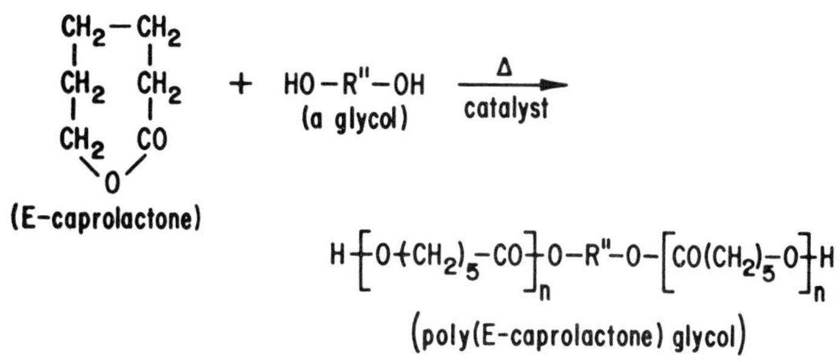

Figure 29. Poly(Caprolactone) Glycol.

when hydrolysis has been extensive enough to produce appreciable poly (ester-urethane) chain fragments terminating at both ends in carboxyl groups.

Further hydrolysis, beyond the capacity of the added poly(carbodiimide), would destroy this network. We suspect we are seeing this network breakup in the cases of the poly(carbodiimide) compounds of polyurethanes (D) and (G) of Figure 28 somewhere between the sixth and ninth weeks of exposure. But it has not become apparent in the compound of polyurethane (K) with poly(carbodiimide) after nine weeks of exposure.

Stearic acid softened polyurethane (F) appreciably, suggesting that the act of incorporating the acid by milling degraded the polymer. Figure 28 modulus change on exposure shows how strongly the stearic acid catalyzed hydrolytic degradation of the poly(ester-urethane).

(D) Hydrolysis Stability of Poly(Caprolactone)-Based Thermoplastic Poly(Ester-Urethanes).

Polyester macroglycols of special structure can be prepared by polymerizing ϵ-caprolactone. Figure 29 shows the chemistry involved in the preparation of poly(ϵ-caprolactone) glycol [14]. These commercially available macroglycols prove to be useful components for preparing poly(ester-urethanes) [15].

Figure 30 shows the hydrolysis ability of two thermoplastic poly(ester-urethane) elastomers made from two poly(ϵ-caprolactone) glycols. Both polymers also contained MDI and 1,4-butanediol in the polymerization recipes and were made using the random melt polymerization conditions used to make the other polyurethanes of this study. One polymer (L) was made from low acid number poly(caprolactone) glycol of 840 molecular weight. Two moles of MDI, one of the macroglycol, and one of 1,4-butanediol were used.

The other polymer (M) was made from intermediate acid number poly (caprolactone) glycol of ca. 2000 molecular weight. Three moles of MDI, one mole of the macroglycol, and two moles of 1,4-butanediol were used here.

The plots of Figure 30 do not show impressive hydrolysis resistance for either of the thermoplastic poly(ester-urethane) elastomers based on the two poly(ϵ-caprolactone) glycols. We cannot explain this performance. It is not

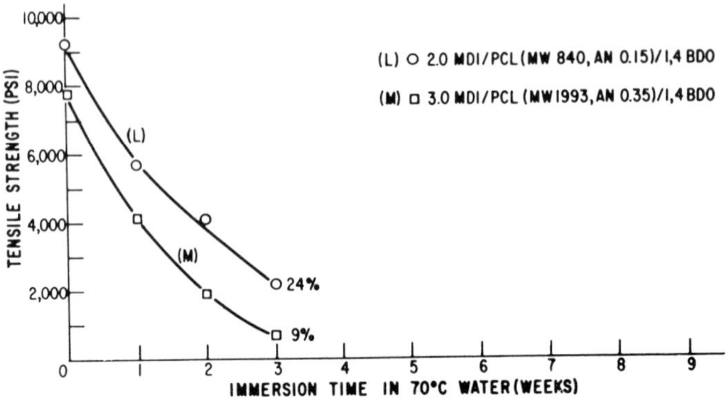

Figure 30. Hydrolysis Resistance of Thermoplastic Poly(Ester-Urethanes) Made From Poly(ϵ-Caprolactone) Glycol of Low Acid Number and Moderate Acid Number.

Macroglycol Name	Poly(ϵ-Caprolactone) Glycol	
Macroglycol Mol. Wt.	840	1993
Macroglycol Acid No.	0.15	0.35
Polyurethane	(L)	(M)
Polyurethane T_2 (°C)	155	160
Polyurethane Hardness (Shore)	89A	89A
Polyurethane Tensile (psi)	9200 (24)*	7800 (9)
Polyurethane Elongation (%)	465 (147)	575 (64)
Polyurethane 300% Modulus (psi)	1500 (73)	1200 (54)
Polyurethane Acid No.	0.21 (1470)	0.42 (1120)

* Number is original value: number in parentheses is % of original value retained on three week immersion in 70°C water.

Figure 31. Hydrolysis Stability of Thermoplastic Poly(Ester-Urethane) Elastomers Based on Poly(ϵ-Caprolactone) Glycol. 2.0 & 3.0 MDI/PCL/1,4-BDO.

consistent with that expected from the relatively low acid numbers and high methylene group/ester group ratios of the poly(ϵ-caprolactone) glycols tested.

However, co-workers at our Avon Lake Development Center have obtained more recent results than these with samples of poly(ϵ-caprolactone) glycol from the same commercial source. In this case the supplier indicated that the macroglycol had been stabilized. The Avon Lake results showed improved hydrolysis stability in typical poly(ester-urethanes) derived from the stabilized poly(ϵ-caprolactone) glycol components.

Figure 31 gives further information on the polymers of Figure 30 and the exposure effects. The initial polymer properties are seen to be quite good. Poly(carbodiimide) stabilization would be expected to improve the hydrolysis resistance of these polymers, as it is reported to do in covalently crosslinked poly(caprolactone)-based poly(ester-urethanes) [9].

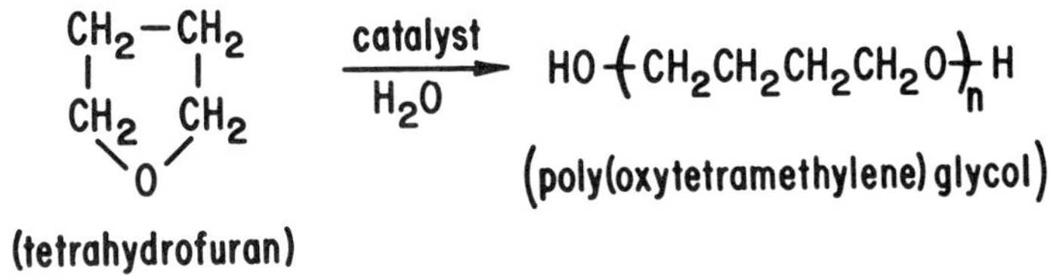

Figure 32. Poly(Oxytetramethylene) Glycol.

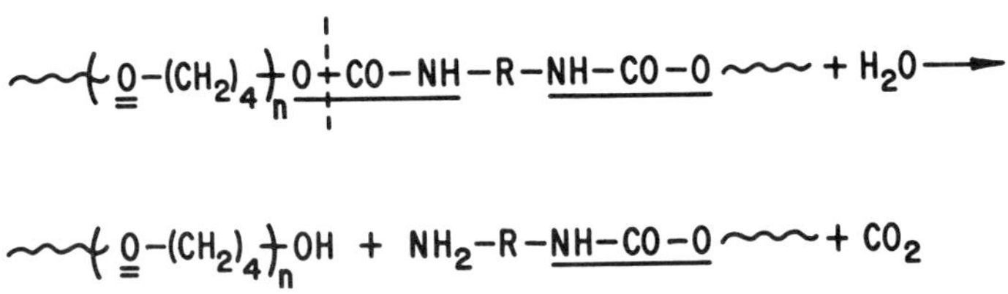

Figure 33. Poly(Ether-Urethane) Structure and Hydrolysis.

(E) Hydrolysis Stability of Thermoplastic Poly(Ether-Urethanes)

(1) Polymers Based on Poly(Oxytetramethylene) Glycol.

Polyether glycols are an important class of macroglycols which have found great use as polyurethane components. One commercially available type, poly(oxytetramethylene) glycol, yields particularly high propertied, solid, poly(ether-urethane) elastomers.

The preparation and structure of poly(oxytetramethylene) glycol is shown in Figure 32. The macroglycol is prepared by the polymerization of tetrahydrofuran, and apparently treatment with water [16].

Figure 33 shows the structure of a poly(ether-urethane) made from poly (oxytetramethylene) glycol. We will see that poly(ether-urethanes) are less easily hydrolyzed than most poly(ester-urethanes). This is because the weakest poly(ether-urethane) linkage towards hydrolysis, the urethane linkage (singly underlined), is stronger than the carboxylic ester linkage of most poly (ester-urethanes). The ether linkage (doubly underlined) is very resistant to hydrolysis and hydrolyzed only under special forcing conditions. So in normal service the dominant reaction in the hydrolytic degradation of poly(ether-urethanes) is cleavage of the urethane group. As shown in Figure 33 this reaction breaks the chain producing two smaller chains. One of these shorter chains now terminates at one end in a hydroxyl group (—OH). The other terminates at one end in an amino group ($-NH_2$). Neither of these types of terminal groups appears to autocatalyze the hydrolysis of poly(ether-urethanes) based on aromatic diisocyanates.

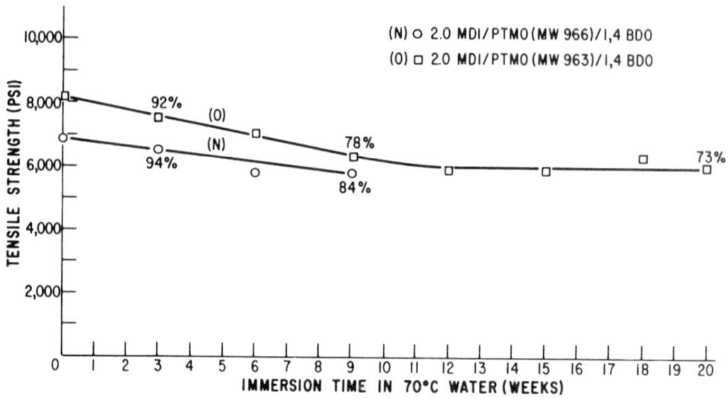

Figure 34. Hydrolysis Resistance of Thermoplastic Poly(Ester-Urethanes) Made From Poly(Oxytetramethylene) Glycol.

Macroglycol Name	Poly(Oxytetramethylene) Glycol	
Macroglycol Mol. Wt.	966	963
Polyurethane	(N)	(O)
Polyurethane DSV	0.797	—
Polyurethane T_2 (°C)	123	144
Polyurethane Hardness (Shore)	86A	89A
Polyurethane Tensile (psi)	6900 (84)*	8200 (77)
Polyurethane Elongation (%)	710 (114)	550 (118)
Polyurethane 300% Modulus (psi)	1200 (110)	1150 (104)

* Number is original value: number in parentheses is % of original value retained on *nine* week immersion in 70°C water.

Figure 35. Changes in Stress-Strain Properties of Poly(Oxtetramethylene) Glycol-Based Thermoplastic Poly(Ether-Urethanes) on Hydrolysis. 2.0 MDI/PTMO/1,4-BDO.

Figure 34 shows the hydrolysis stability of two representative thermoplastic poly(ether-urethane) elastomer samples made from two similar poly (oxytetramethylene) glycols of molecular weight about 1000. In each case one mole of the macroglycols was terpolymerized with two moles of MDI and one mole of 1,4-butanediol. The same random melt polymerization conditions employed to prepare the other polyurethanes of this study were used here.

The outstanding hydrolysis resistance of poly(oxytetramethylene) glycol-based polyurethanes is apparent in the performance of polyurethanes (N) and (O) in the 70°C water immersion test. Note that it was necessary to use an extended time scale for these polymers. Tensile strength retention was 92-94% of the original values after three weeks exposure and 78-84% after nine weeks exposure for these polymers.

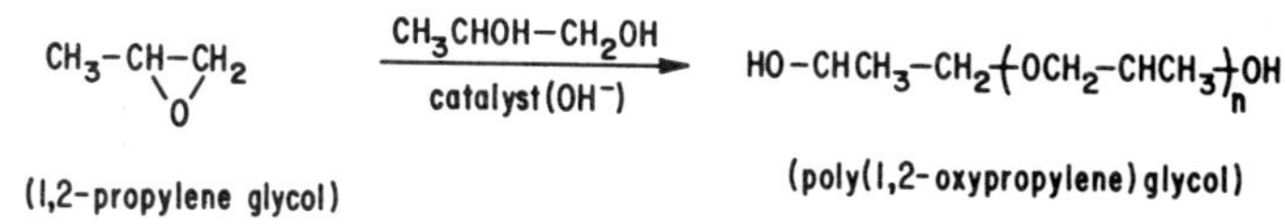

Figure 36. Poly(1,2-Oxypropylene) Glycol.

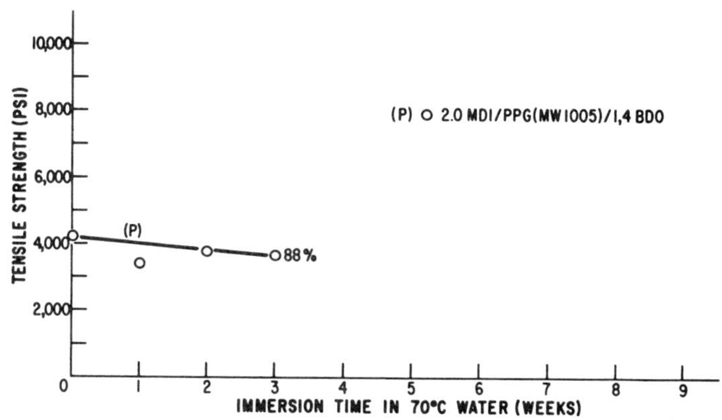

Figure 37. Hydrolysis Resistance of Thermoplastic Poly(Ether-Urethane) Made From Poly(1,2-Oxypropylene) Glycol.

Notice that from twelve weeks of exposure up to twenty-one weeks polyurethane (O) lost essentially no tensile strength. It still had 73% of its original tensile strength after twenty-one week immersion in 70°C water.

Figure 35 provides additional information on the polymers of Figure 34 and the effects of 70°C water immersion. Note that property retention after nine weeks exposure is given in the case of these hydrolysis-resistant poly (ether-urethanes). It is seen that there were no losses in the ultimate elongation or 300% modulus values of polyurethanes (N) and (O) during the nine week exposure.

Poly(1,2-oxypropylene) glycol is another type of polyether macroglycol used to make poly(ether-urethanes) it finds particular use in urethane foam products and has also been used to make solid polyurethane elastomers. Figure 36 shows the preparation of poly(1,2-oxypropylene) glycol by the base-catalyzed polymerization of 1,2-propylene oxide in the presence of some 1,2-propylene glycol [12].

Figure 37 shows the hydrolysis stability of a representative thermoplastic poly(ether-urethane) elastomer sample made by the random melt polymerization of one mole of poly(1,2-oxypropylene) glycol of 1005 molecular weight, two moles of MDI, and one mole of 1,4-butanediol. This polyurethane, (P), shows excellent hydrolysis stability, as expected. After three week immersion in 70°C water it still retains 88% of its original tensile strength.

Macroglycol Name	Poly(1,2-Oxypropylene) Glycol
Macroglycol Mol. Wt.	1005
Polyurethane	(P)
Polyurethane DSV	0.874 (97)*
Polyurethane T_2 (°C)	146
Polyurethane Hardness (Shore)	76A
Polyurethane Tensile (psi)	4200 (88)
Polyurethane Elongation (%)	640 (112)
Polyurethane 300% Modulus (psi)	800 (100)

* Number is original value: number in parentheses is % of original value retained on three week immersion in 70°C water.

Figure 38. Hydrolysis Stability of Thermoplastic Poly(Ether-Urethane) Elastomers Based on Poly(1,2-Oxypropylene) Glycol. 2.0 MDI/PPG/1,4-BDO.

Figure 38 gives further information on poly(ether-urethane) (P) and the exposure effects. The polymer loses no elongation or 300% modulus during three week immersion at 70°C.

In summary this paper has reviewed the types of monomers usually used to make thermoplastic polyurethane elastomers, how they incorporate in the polymer structure, and the unique properties these polymers have.

In discussing the hydrolysis stability of thermoplastic polyurethane elastomers the suitability of these polymers for such study was pointed out and the reasons for choosing a total immersion test explained.

A relationship between poly(ester-urethane) composition in terms of methylene group concentration and chain stiffness, and hydrolysis stability was shown.

The dominant role of polyurethane acid number in thermoplastic poly (ester-urethane) hydrolysis stability was demonstrated and the origin of this unexpected acid number was discussed.

The pronounced stabilizing action of added poly(carbodiimide) in thermoplastic poly(ester-urethane) hydrolysis was shown. Evidence for the mending action of poly(carbodiimide) on broken poly(ester-urethane) chains was presented.

The severe de-stabilizing action of an added carboxylic acid commonly used as a lubricant in rubbers and plastics in the hydrolysis of thermoplastic poly(ester-urethanes) was demonstrated.

The hydrolysis stability of thermoplastic poly(ester-urethane) elastomers based on poly(ϵ-caprolactone) glycol was described. It was shown that in this case low acid number and high methylene group concentration which usually enhance hydrolysis stability were not effective stabilization factors.

Finally, the superior, inherent hydrolysis stability of thermoplastic poly (ether-urethanes) based on two types of commercial polyether glycols was described.

REFERENCES

1. C. S. Schollenberger, H. Scott, G. R. Moore, *Rubber World, 137,* (4), pp 549-555, Jan., 1958. Also, *Rubber Chemistry & Technology, 35,* (3), July-Sept., 1962, pp 742-752.
2. C. S. Schollenberger and K. Dinbergs, SPE Transactions, *1,* (1), pp 31-40, Jan., 1961.
3. W. Cooper, R. W. Pearson, S. Darke, *The Industrial Chemist, 36,* 121-126, (1960).
4. W. Neumann, J. Peter, H. Holtschmidt, W. Kallert, Proceedings of the 4th Rubber Technology Conference, London, 1962, (59); see *Chemical Abstracts, 60,* 9456, (1964).
5. W. Neumann, H. Holtschmidt, J. Peter, P. Fischer, "Stabilization of Polyesters with Polycarbodiimide," U. S. Patent 3, 193, 522, Farbenfabriken Bayer, July 6, 1965.
6. R. J. Athey, Rubber Age (N.Y.), *96,* (5), 705, (1965).
7. Z. T. Ossefort and F. B. Testroet, "Hydrolytic Stability of Urethane Elastomers," 89th Meeting, Rubber Division, American Chemical Society, San Francisco, Cal., May 4, 1966. Also, *Rubber Chemistry & Technology, 39,* (4), Part 2, Sept., 1966, pp 1308-1327.
8. F. H. Gahimer and F. W. Nieske, "Hydrolytic Stability of Urethane and Polyacrylate Elastomers in Humid Environments," *J. Elastoplastics, 1,* Oct., 1969, p.266.
9. G. Magnus, R. A. Dunleavy, F. E. Critchfield, "Stability of Urethane Elastomers in Water, Dry Air, and Moist Air Environments," Rubber Division Meeting, American Chemical Society, San Francisco, May 3-6, 1966. Also, *Rubber Chemistry & Technology, 39,* (4), Part 2, Sept., 1966, pp 1328-1337.
10. F. D. Stewart, "Hydrolysis-Resistant Poly(Ester-Urethanes)," U. S. Patent 3, 463, 758, B. F. Goodrich Co., Aug. 26, 1969.
11. W. Dieckmann and F. Breest, *Berichte, 39,* 3052, (1906).
12. J. H. Saunders and K. C. Frisch, "Polyurethanes—Chemistry and Technology," *(High Polymers, XVI,* Part I), Interscience, (1962).
13. M. Morton, M. A. Deisz, M. Ohta, "Degradation Studies on Condensation Polymers." U. S. Dept. of Commerce Report PB-131, 795, March 31, 1957.
14. R. A. Dunleavy, F. E. Critchfield, "Polyurethane Elastomers: Influence of Structure on High and Low Temperature Properties," *Rubber World, 156,* (3), pp 53-57, June, 1967.
15. G. Magnus, "Poly(ϵ-Caprolactone)-Based Urethanes," *Rubber Age, 97,* (4), pp 86-93, July 1965.
16. Quaker Oats Co., Technical Bulletin #150, Aug. 7, 1964.
17. C. Naegli and A. Tyajbi, Helv. Chim. Acta, *17,* 931, (1934).

C. S. SCHOLLENBERGER and F. D. STEWART, *The B. F. Goodrich Company, Research Center, Brecksville, Ohio*

CHAPTER 5

EFFECT OF CHEMICAL STRUCTURE ON THE THERMAL STABILITY OF CROSSLINKED URETHANE ELASTOMERS

INTRODUCTION

This paper presents studies that were designed to gain insight into the effect of chemical structure on the thermal stability of polyurethane elastomers. The kinetic and thermodynamic parameters associated with urethane cleavage reaction were assessed by means of chemical stress relaxation measurements on specially synthesized urethane elastomer networks. The need to use these special networks should become clear from the background material developed early in the paper. The first part of this paper describes how the stress relaxation behavior of these networks was positively related to the urethane cleavage process and touches on the nature of the urethane cleavage reaction itself. The second part not only describes the effect of diisocyanate structure on the kinetic and thermodynamic parameters associated with the urethane cleavage reaction; it also shows how the nature of the soft segment, as for example substitution of a polyether for a polyester glycol in the urethane backbone, affects the nature and rate of thermal breakdown of the urethane elastomer networks.

STRESS RELAXATION TECHNIQUE & KINETICS OF URETHANE CLEAVAGE REACTION

The techniques of continuous and intermittent stress relaxation measurement have been used successfully for studying the thermal behavior of a wide variety of elastomers [1-9]. Briefly, with continuous stress relaxation a strip specimen of an elastomer is held at a fixed extension while it is being exposed to a constant elevated temperature. Meanwhile, the tensile force or stress at constant extension and temperature is monitored continuously for an extended period of time. If the temperature is high enough to induce random thermal scission of the elastomer's network chains, the measured force, or stress will decrease with time at a rate which is proportional to the rate of chain scission. Such a stress decay process is often referred to as a chemorheological or, more simply, a chemical stress relaxation.

Although the continuous stress relaxation technique can provide a direct measure of the rate of network chain scission, with it one is unable to tell whether or not or to what extent the chain scission reaction is reversible. This is because any recombination of thermally cleaved chains that might occur during a continuous stress relaxation measurement takes place in a relaxed condition with respect to the geometrical constraints on the specimen. It is for this reason that continuous relaxation measurements are often supplemented by intermittent relaxation measurements. In the latter case, the specimen is maintained in a relaxed or unstretched condition at the same constant elevated temperature. However, at widely spaced intervals of time the elastomer strip is stretched momentarily to a fixed extension, the resulting stress is rapidly measured, after which the strip is immediately returned to its unstretched state. Under these intermittent relaxation conditions, virtually all the chain scissions and recombinations are thought of as occurring while the elastomer strip is in the relaxed state. This measurement then follows the net effect of chain scission and recombination with time of heat aging. Since continuous stress relaxation measurement follows the rate of chain scission alone, a combination of continuous and intermittent measurements enables us to assess the relative rates of chain scission and recombination, and hence provides us with a measure of the reversibility of a given chain scission reaction.

To illustrate this point, if the thermally-induced chain scission reaction is completely irreversible, both the continuous and intermittent relaxation curves we would obtain would decay to zero, identically. If, however, the chain scission reaction is completely reversible, that is to say at any given time the rate of chain scission exactly equals the rate of recombination, we would find that the continuous relaxation curve would decay to zero, but the intermittent one would not decay at all.

For dealing with the chemorheological stress relaxation of elastomers, the continuous stress relaxation data are often fitted to a simple exponential model, known as the Maxwell model, [1, 2, 10_a] expressed mathematically as follows:

$$f(t) = f(o)\ \mathrm{EXP}\ (-t/\tau)$$
$$E_r(t) = E_r(o)\ \mathrm{EXP}\ (-t/\tau)$$

In the first equation, f(o) is the initial tensile force measured at constant extension and constant temperature, and f(t) is the tensile force measured after some relaxation time t. The second equation is an alternate expression for the Maxwell model in terms of the relaxation moduli $E_r(o)$ and $E_r(t)$. These modulus values are directly calculable from the corresponding force values, the cross-sectional area of the specimen, and the extent of its stretch (10b). The constant τ in these equations is a characteristic relaxation time constant, which is temperature dependent in a manner reminiscent of a chemical reaction rate constant. When applying either of these equations to the chemical stress relaxation of elastomer networks undergoing random thermal cleavage, one assumes that the measured force or stress at any time is proportional to the number of stress bearing network chains that have never been cut up to that time. It is then possible to relate the relaxation time constant τ to the chain scission reaction rate constant k. More specifically, $1/\tau = k$.

Turning to the polyurethane elastomers, it has been generally accepted that the thermal cleavage of the urethane linkages present in the network chains is responsible for the thermal degradation of these elastomers when subjected to temperatures of 100° to 200°C. With this thought in mind various investigators [11-18] used the stress relaxation technique to explore the kinetics of urethane cleavage in this temperature range. If one assumes the network chain scission to be restricted to the urethane linkages, the relationship between the relaxation time constant τ and the chain scission reaction rate constant k should contain a parameter, say m, which represents the number of urethane linkages contained in a network chain. In fact, it can be shown from theoretical considerations [13, 18] that for the case of polyurethane networks $1/\tau$ should equal km.

Now, to determine the specific reaction rate constant k for urethane cleavage, as well as the thermodynamic constants associated with this reaction by means of stress relaxation measurements, it became necessary to develop a technique for synthesizing clean and well defined urethane networks for which m, the number of urethane linkages per network chain would be fixed and known precisely. To this end, model isocyanate-terminated urethane prepolymers were prepared from an aromatic diisocyanate and a hydroxy-terminated polyester. The molecular structure of this model prepolymer consists of one molecule of the hydroxy-terminated polyester which has been reacted on either end with one molecule of the aromatic diisocyanate. In this case the diisocyanate used is 2,4-toluenediisocyanate:

The details of preparing these model prepolymers as well as the proof of their structure, has already been published elsewhere [18]. Suffice it to say that in preparing these model prepolymers four major side reactions had to be avoided. These were the reaction of the NCO groups with either moisture or free acid end groups of the polyester glycol to produce urea linkages and biuret branch points; the reaction of NCO groups with urethane groups of the prepolymer to form allophanate branch points; and chain extension of the prepolymer through reaction of its terminal NCO groups with unreacted glycol to give higher molecular weight prepolymer species. The first two side reactions were avoided or at least minimized by working under anhydrous conditions and using polyester glycols with as low an acid number as possible. The third reaction, that of allophanate formation, could be minimized by maintaining the reaction temperature at or below 50°C. Finally and most importantly, one could suppress prepolymer chain extension by reacting the polyester glycol with a large excess of diisocyanate and then quantitatively stripping off the excess diisocyanate at the end of the reaction through continuous liquid/liquid extraction using hexane as the extraction solvent.

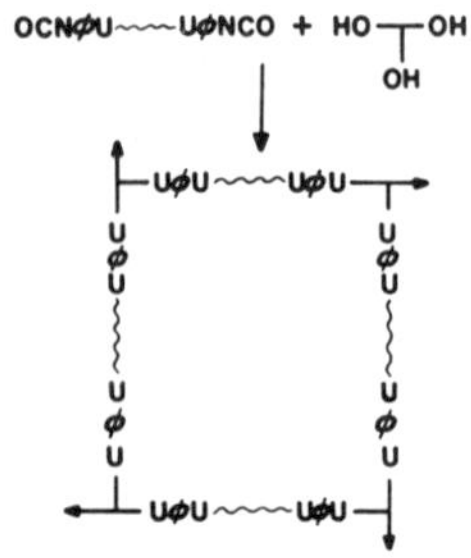

Figure 1. Model Urethane Network Formation.

When these NCO terminated prepolymers are reacted with an equivalent (based on the NCO content) of a simple triol, such as 1,1,1-trimethylolpropane, one obtains at complete reaction, a highly uniform network as shown in Figure 1. Each network chain of such model networks should contain precisely four urethane linkages.

Now, if such polyurethane networks were free from extraneous weak linkages, and if then the urethane linkages were the only thermally cleavable ones present, the chemical stress decay behavior of these elastomers should be governed by the simple Maxwellian rate expression described earlier with km replacing $1/\tau$. From the simple exponential nature of this equation, we would expect semi-log plots of the stress decay data for these elastomers to be linear with slopes equal to -km/2.3, and with m, of course, equal to 4.

Experimentally, these elastomers, fabricated in the form of sheets 75 mils thick, were die cut into suitable T-50-type specimens. Then specimens were thoroughly extracted with tetrahydrofuran before subjecting them to stress relaxation measurement to remove all traces of catalysts or other extractable impurities that might adversely affect their thermal stability.

The specimens were then subjected to continuous and intermittent stress relaxation measurement on a six-channel autographic stress relaxometer of our own design, which combines high sensitivity with excellent long-term stability.

Figure 2 shows a semi-log plot of the continuous stress relaxation data obtained at 150°C. on one of these specially synthesized urethane networks. This network was prepared by reacting a model prepolymer consisting of a 2,4-TDI capped polyethylene-adipate glycol with an equivalent of trimethylolpropane. Also shown on this plot are the intermittent relaxation data obtained with this network at the same temperature. These curves are typical in form of those obtained with other polyester-based urethane networks at the various experimental temperatures. The continuous stress decay curve is seen to be essentially linear although some initial deviation from linearity is evident. This initial non-linearity was traced to the thermal cleavage of a few random weak

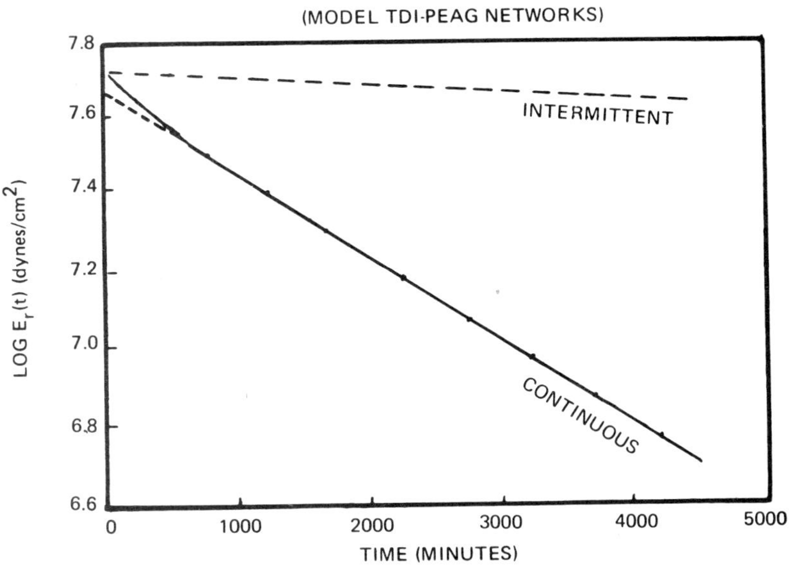

Figure 2. Continuous and Intermittent Stress Decay Plots at 150°C.

linkages of the types mentioned earlier that were formed inadvertently during the network synthesis [18]. The intermittent relaxation plot is essentially invariant with time. This behavior is indicative of reversible chain scission, such as would be expected of the urethane cleavage reaction. However, in order to unequivocally identify this stress relaxation behavior with urethane cleavage we synthesized a series of three model prepolymers from 2,4-TDI and polyethylene-adipate glycols of different number average molecular weight. These model prepolymers were reacted with an equivalent of trimethylolpropane to form urethane networks called E-1, E-2, and E-3. In this series of elastomers the molecular weight of the polyester glycols was varied over a three-fold range, but in all cases the number m of urethane linkages per network chain was held constant at four. Figure 3 shows semi-log plots of the continuous stress decay data obtained with this series at 150°C. As you can see, these plots are all essentially linear. However, the important thing to observe here is that the slopes of the linear portions of these plots are all equivalent. This result is quite consistent with the theoretical equation for stress decay due to urethane cleavage, since these networks are all chemically similar and they all contain 4 urethane groups per network chain.

Now, in order to determine the kinetic and thermodynamic constants associated with the urethane cleavage reaction, one simply calculates values of k from the slopes of semi-log plots of continuous stress decay data taken at various experimental temperatures. For making these calculations m is taken to be four in all cases. Also appearing in this figure is an Arrhenius plot of

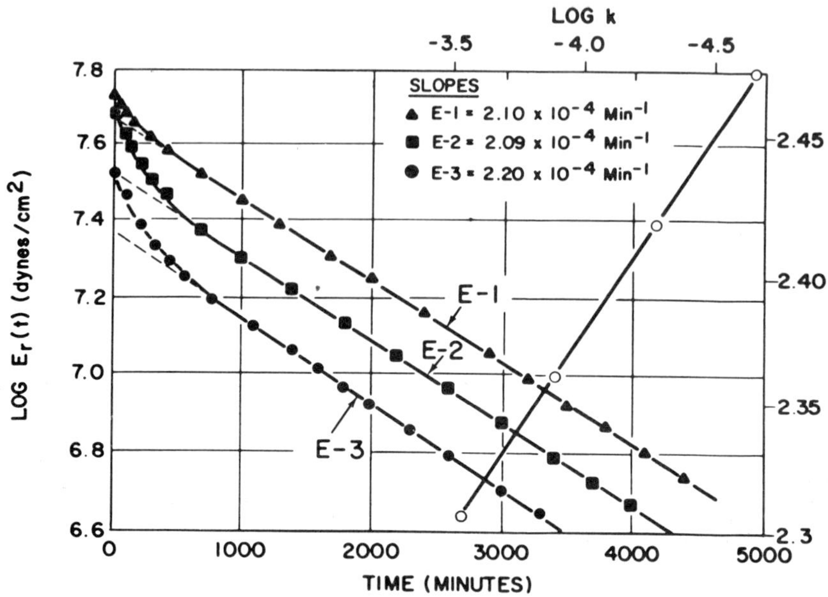

Figure 3. Stress Relaxation Plots of Elastomers E-1, E-2 and E-3 at 150°C.

the k values at various temperatures. The energy of activation for the urethane cleavage reaction was calculated from the slope of this plot.

It should be noted at this point that although elastomers E-1, E-2, and E-3 were all made to contain the same number of urethane linkages per network chain, the concentration of urethane linkages per unit network volume was varied by almost threefold in this series. This variation in urethane linkage concentration per unit network volume was brought about through the use of polyester glycols of different chain lengths in preparing the model prepolymers. It is apparent from data on E-1, E-2, and E-3 that the stress decay rate, which can now be associated with urethane cleavage, is independent of the concentration of urethane linkages per unit network volume, at least over this threefold concentration range. This observation, together with the obvious linearity of the semi-log plots, strongly suggest that the urethane cleavage reaction follows a first order rate law, or at least an apparent first order law.

THE EFFECT OF DIISOCYANATE STRUCTURE ON THE THERMAL BEHAVIOR OF POLYURETHANES

In order to gain insight into the effect of diisocyanate structure on the thermal behavior of the urethane linkages derived from them, a series of diisocyanates was selected. These diisocyanates, shown in Table 1 were selected on the basis of their academic and commercial interests.

Each of these diisocyanates was reacted with a 2,000 Mn polyethylene adipate glycol under the conditions described earlier to produce model NCO-terminated prepolymers. These model prepolymers were then reacted with an equivalent of trimethylolpropane to form elastomer networks whose network chains all contain precisely four urethane groups. After thorough solvent extraction, suitable specimens of these elastomers were subjected to continuous stress relaxation measurement at four different temperatures between 130° and 160°C. From semi-log plots of these stress decay data values for the specific reaction rate constant k and the molar energy and entropy of activation were calculated for the thermal cleavage reaction of each type of urethane linkage by the procedure outlined earlier.

On Table 2 [9] are summarized the thermal cleavage rate data for the series of urethane linkages based on MDI, DMMDI, and TODI. Here we see that the specific rate constant for the cleavage of urethane linkages based on DMMDI and TODI at any given temperature is about twice as large as that for the one based on MDI. The distinguishing structural feature of the DMMDI- and TODI-based urethane linkages relative to that based on MDI is the presence of the methyl group in the ring position ortho to the urethane linkage. Theoretically, this methyl group exerts two opposing effects on the stability of the urethane linkage. One is an electronic inductive effect, which tends to reduce the lability of the proton on the urethane nitrogen, thereby tending to stabilize the urethane linkage. The other is a steric or spacial crowding effect in the vicinity of the urethane linkage which tends to destabilize this linkage. The experimental results point to the overwhelming influence of the steric effect of the ortho-methyl group over its electronic inductive effect. This steric effect is also evidenced by the smaller negative entropy of activation associated with the cleavage of urethane linkages based on DMMDI and TODI relative to that associated with the one based on MDI. One further point worthy of mention is that the presence of a methylene bridge between the aromatic rings of the diisocyanate has little if any effect on the thermal behavior of the urethane

Table 1. Diisocyanates.

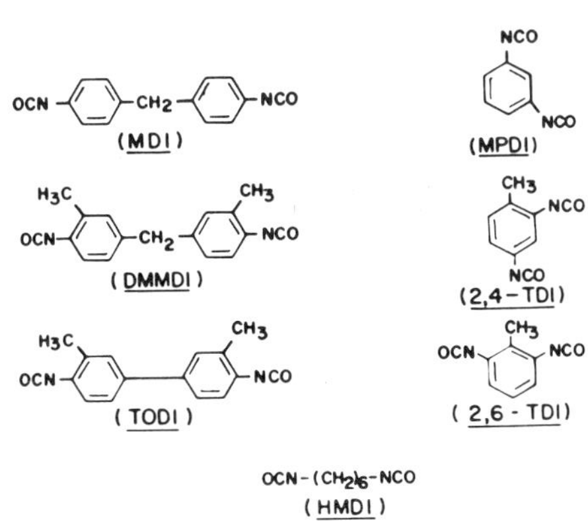

Table 2. Kinetic and Thermodynamic Constants for Urethane* Cleavage.

DIIOCYANATES	$k \times 10^5$ Mn^{-1} 130°C	140°C	150°C	160°C	ΔE* k cal/mole	ΔS* e.u./mole
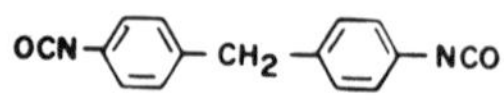 (MDI)	1.69	5.75	12.0	27.9	31.9 ± 1	-12.5 ± 2
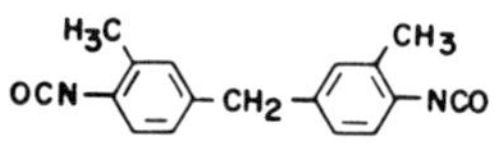 (DMMDI)	–	9.83	27.3	60.4	31.3±1	-7.2±2
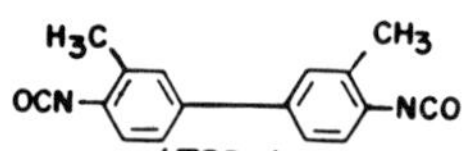 (TODI)	3.34	8.4	25.6	58.0	33.1±1	-8.3±2

*URETHANE BASED ON PEAG AND DIISOCYANATE DESIGNATED ABOVE.

Table 3. Kinetic and Thermodynamic Constants for Urethane* Cleavage.

DIISOCYANATE	$k \times 10^5$ min^{-1} 130°C	140°C	150°C	160°C	ΔE* k cal/mole	ΔS* e.u./mole
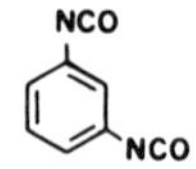 (MPDI)	2.18	5.64	14.10	28.80	30.0±1	-16.3±2
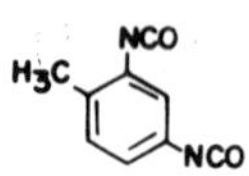 (2,4-TDI)	2.09	4.73	12.30	28.60	29.6±1	-17.10±2
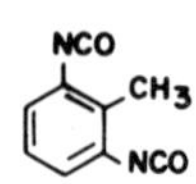 (2,6-TDI)	4.55	12.5	34.2	84.0	34.5±1	-2.46±2

*URETHANE BASED ON PEAG AND DIISOCYANATE DESIGNATED ABOVE.

linkages. This point is evident from a comparison of the data obtained with the DMMDI- and TODI-based networks.

The data in Table 3 summarizes the results obtained with the series of urethane networks based on *meta*-phenylenediisocyanate or MPDI, 2,4-TDI and 2,6-TDI. The higher specific rate constants and smaller negative entropy of activation associated with the thermal cleavage of the urethane linkage based on 2,6-TDI relative to those values obtained for the cleavage of the urethane linkage based on 2,6-TDI relative to those values obtained for the cleavage of the urethane linkage based on MPDI again point to the adverse effect on thermal stability of the ortho-methyl group through steric crowding. The higher energy of activation observed in the latter case could be due to the inductive effect of the methyl group on the two adjacent urethane linkages. The kinetic picture in the case of the urethane network based on 2,4-TDI is somewhat more complex. Here, two of the four urethane linkages contained in each network chain are joined to the TDI ring at the 2-position, the other two being joined at the 4 position. One would, therefore, expect the observed specific cleavage rate constant at any given temperature to be the average of the specific rate constants for the two types of linkages. One would further expect this average value to be somewhere between those observed with the MPDI- and 2,6-TDI-based networks. In fact, we find that it lies close to that of the more stable MPDI-based network. A possible explanation for this result is the stabilizing influence of the methyl group on the urethane linkage at the 4-position due to its electronic inductive effect.

Chemical stress relaxation measurements on the urethane network based on an aliphatic diisocyanate, hexamethylenediisocyanate, indicated a complex behavior. It was found that the stress decay exhibited by this urethane network at 130°C. is governed mainly by an irreversible oxidative cleavage reaction. This point is illustrated in Figure 4 in which are plotted the continuous and intermittent relaxation data obtained with this network in air as well as intermittent relaxation data obtained using a nitrogen atmosphere.

The decrease in relative modulus as measured intermittently in air shows that the network chain scission process involved is largely irreversible in this environment. The fact that the relative modulus measured intermittently in a nitrogen atmosphere shows little change, provides conclusive evidence that the irreversible network cleavage reaction is oxidative in nature.

THE EFFECT OF SOFT SEGMENT ON THE THERMAL BEHAVIOR OF POLYURETHANE ELASTOMERS

In the stress relaxation studies described thus far, the urethane networks involved all had a common soft segment derived from polyethyleneadipate glycol. In order to explore the effect of variations in soft segment structure on the thermal behavior of polyurethanes, two additional 2,4-TDI-based networks were prepared by our special synthesis techniques. One was made from a model

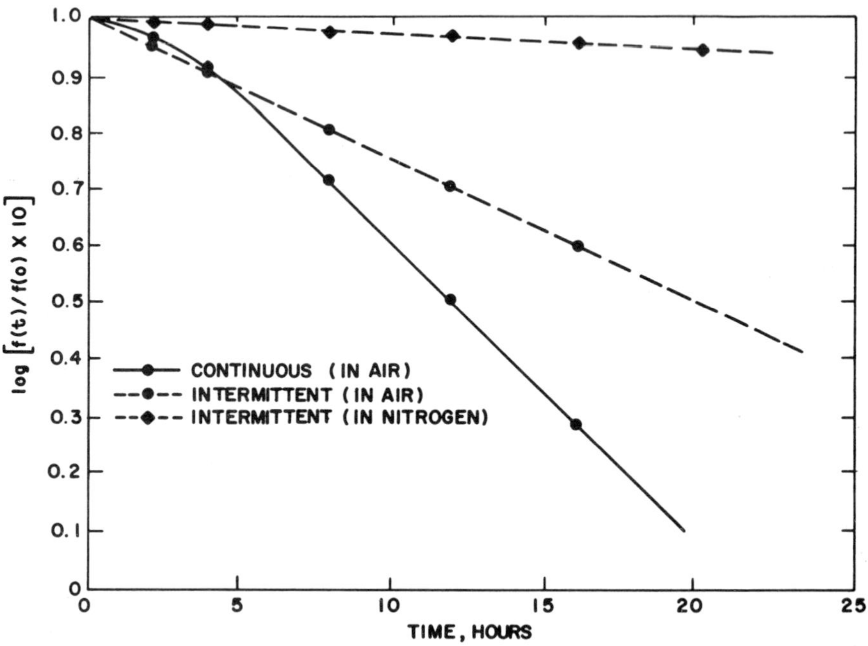

Figure 4. Stress Decay Plots for HMDI-Based Networks at 130°C.

prepolymer derived from polybutyleneadipate glycol, while the other was made from a model prepolymer derived from polytetramethylene ether glycol. Both the polyester and polyether glycols used had a number average molecular weight of 2,000. The model prepolymers were both reacted with an equivalent of trimethylolpropane to form the desired networks, which were thoroughly solvent extracted prior to being used.

Figure 5 shows a comparison of the stress relaxation behavior, in this case at 160°C., of the urethane network derived from 2,4-TDI and polybutylene adipate glycol with that of a corresponding network derived from 2,4-TDI and polyethylene adipate glycol. The similarity in slope of the linear portion of these stress decay curves shows that these polyester glycol moieties can be interchanged without significantly affecting the thermal stability of the urethane linkages derived from them.

Figure 6 shows a comparison of the thermal behavior of the polyether-based network derived from polytetramethylene ether glycol and 2.4-TDI with that of the corresponding polyester-based one derived from polyethylene adipate glycol and 2,4-TDI. These data were obtained at 130°C. A comparison of the continuous stress relaxation curves clearly indicates that the rate of chain scission at 130°C. is considerably greater for the polyether-based network than it is for the polyester-based one. The fact that the intermittent relaxation curve of the polyether-based network decays almost as rapidly as the continuous

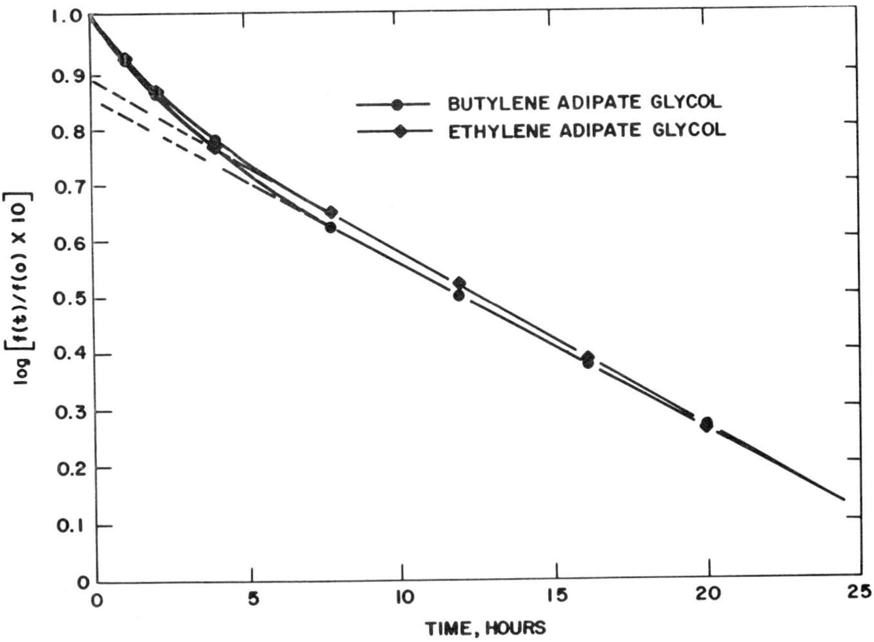

Figure 5. Stress Decay Plots of TDI-Based Networks at 160°C.

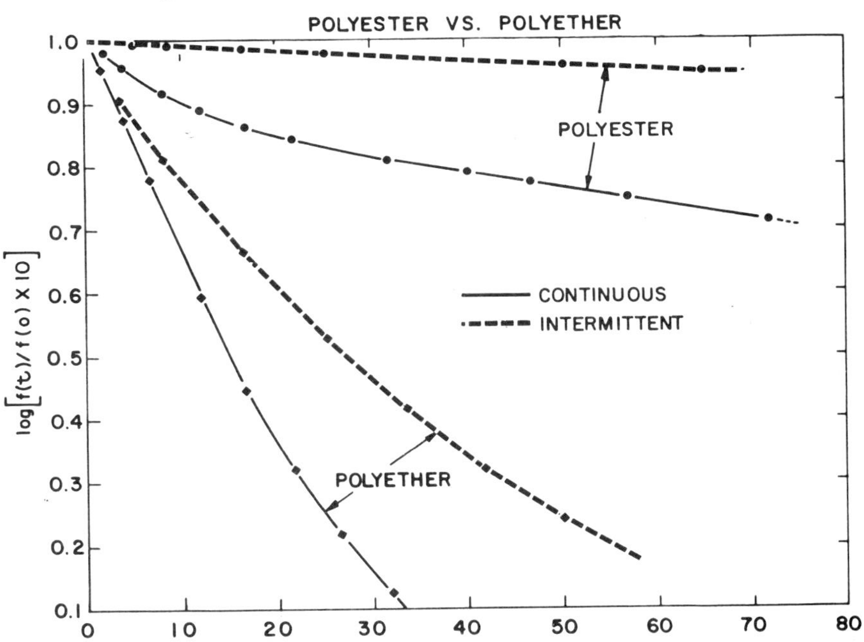

Figure 6. Stress Relaxation Plots at 130°C.

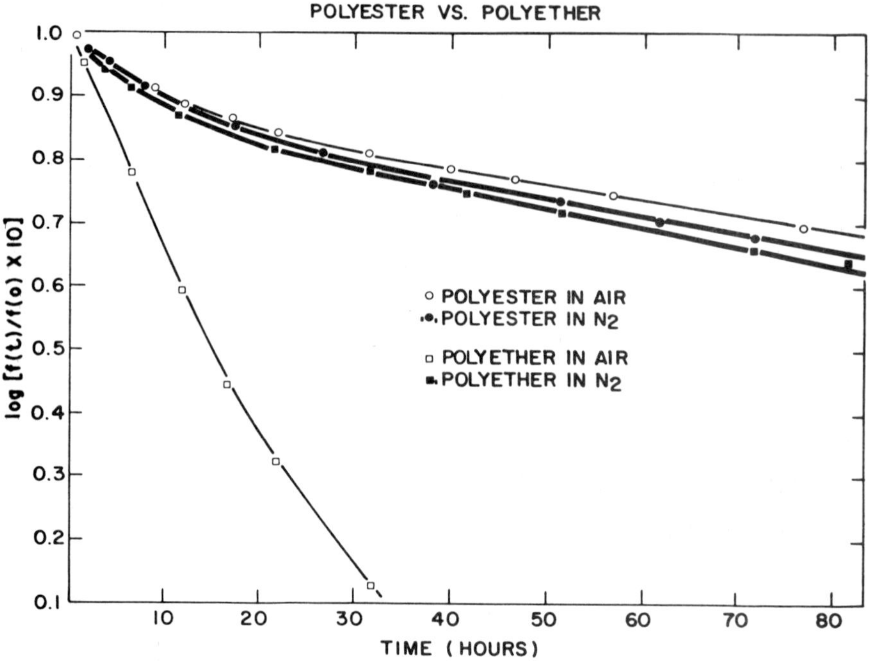

Figure 7. Continuous Stress Relaxation in Air and in Nitrogen at 130°C.

relaxation curve shows that the thermal breakdown of the polyether-based network is dominated by an irreversible process. The data appearing on the next two slides provided conclusive evidence that this dominant, irreversible process involves the oxidative cleavage of the polyether chain segments. A comparison of the continuous stress relaxation data at 130°C. taken in air vs. those taken under nitrogen is shown in Figure 7. The excellent agreement between the curves obtained with the polyester-based network in air and under nitrogen indicates that the chain cleavage processes taking place in this network are non-oxidative. The interesting thing here is that the stress decay rate of the polyether-based network, which is so rapid in the air environment at 130°C., becomes indistinguishable from that of the polyester-based one in nitrogen environment. This result clearly shows that the irreversible chain scission process dominating the stress decay of the polyether-based network is oxidative in nature.

It further shows that the thermal stability of the urethane linkages of the polyether-based network is similar to that of the urethane linkages contained in the corresponding polyester-based network.

A comparison of the intermittent relaxation data at 130°C taken in air vs. those taken under nitrogen is portrayed in Figure 8.

Again one sees that the thermal cleavage taking place in the polyester-

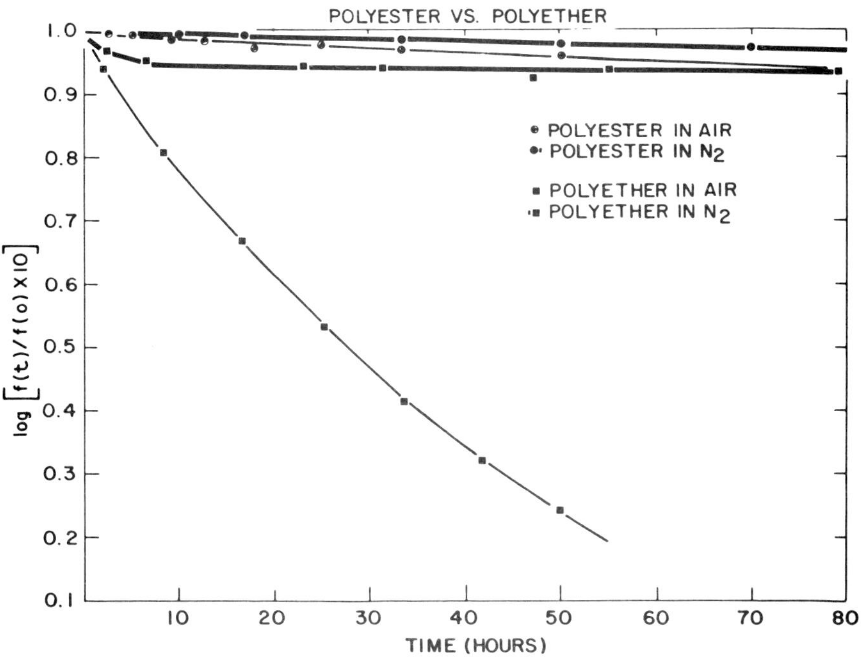

Figure 8. Intermittent Stress Relaxation in Air and in Nitrogen at 130°C.

based network is both reversible and non-oxidative, since the relative modulus of this network as measured intermittently either in air or in nitrogen remains essentially invarient with time of heat aging. Also, as one might expect, comparable intermittent relaxation behavior is observable with the polyether-based network only in the nitrogen atmosphere.

Thus far conclusions as to the relative thermal behavior of the polyester and polyether-based networks have been drawn from stress relaxation data taken at 130°C. Figure 9 summarizes the continuous stress relaxation data we obtained on these networks at various experimental temperatures in the range of 50° to 150°C. Here, the data have been plotted on a log time scale so that they could all be accommodated on the same slide. The effect of oxidative cleavage of polyether chain segments in the polyether-based network is observable at each of the experimental temperatures. Although the measurements at 50°C. are necessarily limited because of the long relaxation times involved, sufficient data have been obtained to show the trend.

It was of interest, in this connection, to determine the effect of antioxidants on the thermal behavior of the polyether-based elastomer. Two of the more common antioxidants were selected for this purpose. One of them was Antioxidant 2246 or 2,2′-methylene bis(4-methyl-6-t-butylphenol), which is representative of the hindered phenol-type, and the other was phenyl-beta-naphthylamine, representative of the amine-type antioxidants. Each of these antioxidants was

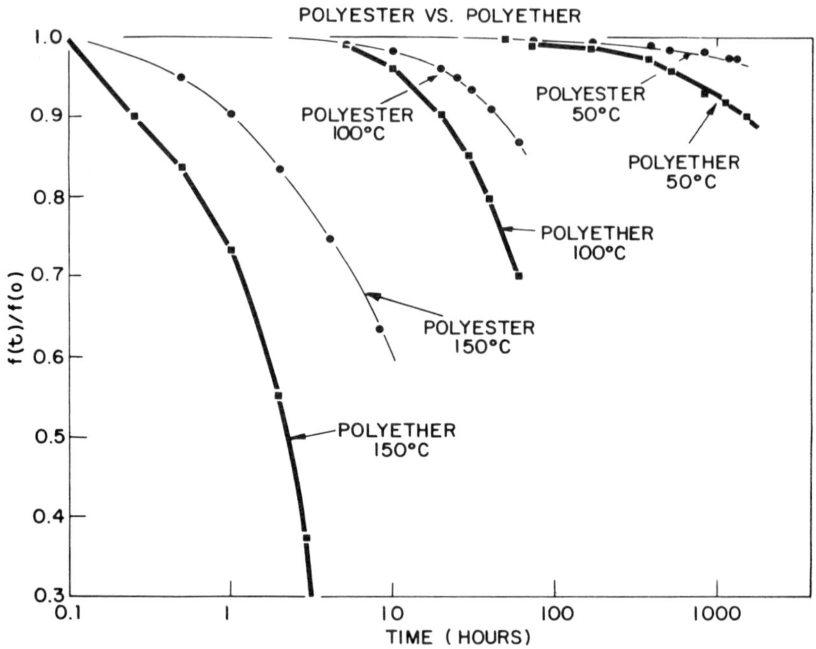

Figure 9. Continuous Stress Relaxation at Various Temperatures.

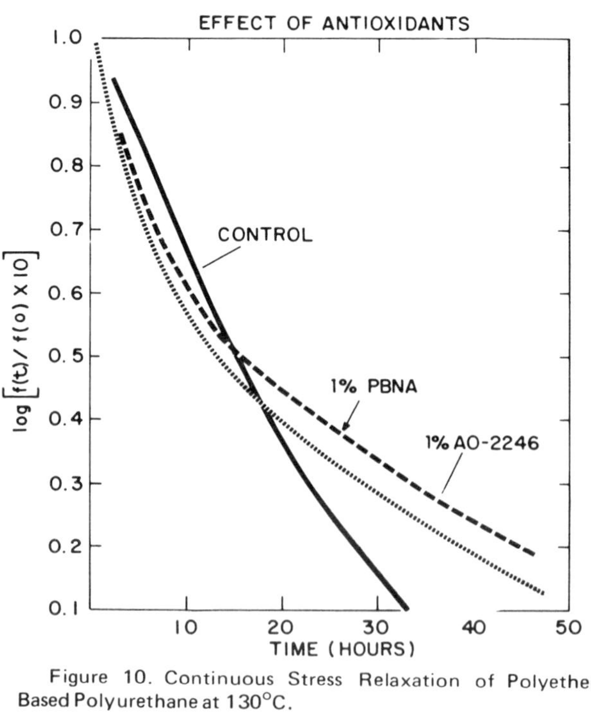

Figure 10. Continuous Stress Relaxation of Polyether Based Polyurethane at 130°C.

added at the 1% level to the model polyether-based prepolymer just prior to its reaction with the triol curative. Figure 10 shows continuous stress relaxation data we obtained at 130°C on the polyether-based network, both with and without the antioxidants. It is apparent that these antioxidants fail to protect the polyether-based urethane, and in fact, they tend to further accelerate its degradation in the initial phase. The probable explanation for these results is that these antioxidants contain chemical groups that have active hydrogens on them. When incorporated in the urethane prepolymers, a portion of the antioxidants could be rendered ineffective through reaction of their active hydrogens with free isocyanate groups of the prepolymers. It is also probable that these antioxidants act as catalysts for the thermal cleavage reaction of urethane linkages or other network linkages and branch points derived from isocyanates.

REFERENCES

1. A. V. Tobolsky, I. B. Prettyman and J. H. Dillon, *J. Appl. Phys.,* 324, 380 (1944); Rubber Chem. Technol., *17,* 551 (1944).
2. R. D. Andrews, A. V. Tobolsky and E. E. Hanson, *J. Appl. Phys. 17,* 352 (1946).
3. R. B. Beevers, *J. Colloid Sci., 19,* 40 (1964).
4. J. Scanlan, *Trans. Faraday Soc., 57,* 839 (1961).
5. L. E. St. Pierre, and A. M. Bueche, *J. Phys. Chem., 63,* 1338 (1959).
6. J. R. Dunn, J. Scanlan and W. F. Watson, *Trans. Faraday Soc.,* 27, 730 (1958).
7. J. P. Berry, *Trans. Inst. Rubber Ind., 32,* 224 (1956).
8. R. C. Osthoff, A. M. Bueche and W. T. Grubb, *J. Am. Chem. Soc., 76,* 4659 (1954).
9. A. V. Tobolsky and A. Mercurio, *J. Polymer Sci.,* 36, 467 (1959).
10. A. V. Tobolsky, Properties and Structure of Polymers, Wiley, New York, 1960 (a) pp 223-265 (b) pp 105-106
11. J. A. Offenbach and A. V. Tobolsky, *J. Colloid Sci., 11,* 39 (1956).
12. P. C. Colodny and A. V. Tobolsky, *J. Am. Chem. Soc., 79,* 4320 (1957).
13. A. V. Tobolsky and E. Peterson, *J. Phys. Chem., 67,* 930 (1963).
14. A. V. Tobolsky, H. Yu and R. Thach, *J. Appl. Polymer Sci., 6,* 44 (1962).
15. A. Singh and L. Weissbein, *J. Polymer Sci.,* A. *3,* 1675 (1965).
16. A. V. Tobolsky, V. Johnson and W. J. MacKnight, *J. Phys. Chem., 69,* 476 (1965).
17. A. V. Tobolsky, *J. Polymer Sci.,* B. *2,* 823 (1964).
18. A. Singh, L. Weissbein, *J. Polymer Sci.,* A. *4,* 2551 (1966).

AJAIB SINGH, *American Cyanamid Company, Bound Brook, New Jersey*

CHAPTER 6

TRENDS IN CAST URETHANE ELASTOMERS

INTRODUCTION

The past decade has been kind to the field of urethane case elastomers. Total volumes of resins used have grown more than ten times. This growth from approximately 3½ million pounds in 1960 to an excess of 50 million pounds in 1970-71 has been made possible by many factors, including the cooperative effort of the manufacturers of the raw materials used in the manufacture of the prepolymers or one-shots, the users of these systems in the development of new casting technology; the machinery manufacturers have developed the high volume accurate mixing-metering equipment; the polymer chemist in designing the polymer to be used, and the design engineer.

The cross fertilization found in the urethane industry has truly been unique. This cross fertilization was necessary because the urethanes were different both in their physical characteristics and in the method of manufacturing of parts. This contact between the end user of parts, manufacturer of parts, and resin supplier was necessary to yield the finished part which would satisfy the requirement for extended service or performance. Without this cross contact, the growth of the field of urethanes would have been much slower because they are basically expensive resins, in comparison to many of the other materials currently in commercial use. Thus, extended durability and performance were and are the major criteria for their use.

The cast urethane industry has evolved through resin and product quality concepts wherein engineering skills or services are as important as the part itself. The high service effort must, obviously, be paid for. Therefore, the industry has remained "fairly high priced" in all facets, but because of the unique physical characteristics or engineering characteristics of the polymer, a sustained growth pattern has been enjoyed and will continue to be enjoyed.

Ten years ago there were six manufacturers of polyurethane prepolymer or one-shot ingredient systems. Today, there are at least eleven major chemical polymer suppliers, and over 85 producers of urethane resins with capacities from 50,000 to 1 million pounds per year, producing specialty resins for specific jobs. Growth in this industry is anticipated to continue at its current 10% per year rate, so that by the end of the next decade, volume in excess of 150 million pounds may be anticipated. In addition to the general market of high strength elastomers, additional volume of material is being used in a rapidly expanding field of flexible molds for the production of such articles as furniture, decorative plaques, or escutcheons, from low density urethane foams. Similar urethane elastomers are being used as pipe gaskets for water and sewage pipes, etc. In addition, potting compounds, which are used to encapsulate electrical components used for advanced design military equipment in aircraft, etc., also make use of these specialty polymers.

"One-shot" polyurethanes are also developing into a growth business. These polymers are being commercially used in the manufacture of rolls, both printing and other, gears, wheels, etc. - in short, many of the articles previously manufactured with prepolymer systems.

DEVELOPMENTS LEADING TO CASTABLE URETHANES

"Polyurethane" elastomers is really a misnomer in that in most of the commercially available resins, the urethane segment of the polymer molecule in fact represents the smallest portion of the total polymer. Perhaps better nomenclature would be polyester or polyether polyamid polyurea polyoallophonate polyurethane. In most instances the polymers that we consider to be polyurethane contain substantial segments of each of the aforementioned chemical groups. In fact, by properly varying reaction conditions and ingredients, the reaction can be driven to yield a preponderance of any one or combination of the groups.

Often this technique is used to change the physical characteristics of a given polymer as is required for the engineering characteristics in the end use article.

Prepolymer polyurethanes, as we know them today, are based upon three building blocks, i.e., isocyanate, polyol, and curative. The "one-shot" systems are based upon isocyanates and a coreactant blend. In both of these systems the structure of each of the components is important in deriving the physical characteristics, as are the reactive sites in determining the final polymer type.

A direct comparison of the change in technology can be made if the separate polymer types of polyether vs. polyester are considered. For example, in polyester technology most of the prepolymer sold in 1960 was based upon the use of either MDI (methylene bis 4-phenyl diisocyanate) or TODI (bitolylenediisocyanate) combined with a polyethylene glycol adipate polyester. The technology at that time was built around the fact that the physical characteristics of the cured polymer could be most readily varied by using a constant prepolymer and a variety of curatives to change polymer structure. This technology did permit the inventory of one or perhaps two prepolymers to produce cured polymers of widely varying characteristics. These polymers generally were high viscosity in nature and fast setting, thus were limited in their ability to be handled at high volume through-put rates. In fact, these polymer systems were most usable in small scale hand-batch open casting or compression molding manufacturing techniques. Further, the use of a fairly regular or symmetrical back-bone such as the polyethylene glycol adipates in the prepolymer, although yielding excellent modulus characteristics, resulted in cured polymers that were deficient in their ability to withstand dynamic stresses as well as in low temperature qualities. In addition, the virtually chemically pure high priced diisocyanates that were required to yield optimum polymer properties foretold that this polymer type would remain relatively high priced, limiting its growth potential in a growth market by virtue of costs.

The polyester-MDI or TODI technology had been a direct carry-over from the Bayer, et al, technology developed in Europe. The labor practices in existence in Europe also lent themselves to the use of these polymers, i.e., hand-batching, small batches, etc., relying on personnel to supply accuracy. U.S. production concepts, however, were and are such that continuous mixing via metering dispensing machines, wherein the reliability is placed upon the machine rather than upon control by production workers, is required. These requirements fostered the change in prepolymer technology to include low viscosity, heat stable prepolymer systems which would be reacted with one curative to yield finished products with the same or greater variety of physical characteristics that were previously obtained with the "European polymer" approach.

At first an attempt was made to utilize MDI and TODI in these prepolymer systems. However, MDI did not prove practical for systems of this type for a number of reasons, including high cost, prepolymer stability, low viscosity polymers (required for mixing-metering machines), and the rate of reaction with available curatives.

Initial attempts at using TDI – as was available in 1960 – also caused some difficulty in that variations in isomer ratio, i.e., 2,4 or 2,6, yielded polymers with varying rates of reactivity with a given curative. In addition, substantial variation in physical characteristics from batch to batch was the reason. Pure 2,4 TDI solved the problem in such a manner that inordinately high plant investment for the prepolymer manufacturers was eliminated.

This approach to prepolymers in this country was initially based primarily upon the use of the PTMG (poly tetramethyleneether glycol) ether. The DuPont Company was the leader in this field, developing both the raw materials based upon their Furan technology, the isocyanate and subsequently the prepolymer system.

The PTMG polyethers were considerably more expensive than the ethers based on polypropylene glycol or polyethylene glycols. However, they offered drastically improved cured polymer characteristics when compared to the cured polymer characteristics based upon the polypropylene or polyethylene glycol systems. The PTMG based polyurethanes offered an unusual degree of hydrolytic stability as well as very high load bearing characteristics even in comparison to many of the then best polyester systems. PTMG ether systems were the first prepolymer systems to incorporate the low viscosity and heat stability characteristics required for large scale machine casting. In this sense the PTMG/TDI prepolymer systems were really the materials that established "castability" that has ultimately led to the rapid growth of the castable urethane elastoplastics markets.

With the PTMG polyether systems establishing the requirements for castability, polyester technology rapidly changed to include polyesters with varying degrees of flexibility, low temperature or high temperature properties. This was generally done by interruption of the crystalline segment of the polyester back-bone by the inclusion of perhaps propylene glycol or butylene glycol. The esters generally then developed into composites of ethylene, propylene, and possibly butylene glycol adipates. Varying the specific ingredients in the polyester back-bone permitted the polyester manufacturer to produce resins which would yield a wider variety of properties in the finished prepolymer and subsequently cured urethane polymer. The technology improved further to yield intermediate polyester resins with substantially lower acid number (thereby eliminating the isocyanate – COOH reaction), thereby reducing the potential of a number of undesirable side reactions in the prepolymer formation. In addition, the polyester manufacturers learned to "control" the molecular weight range (hydroxyl number) so that more uniform prepolymer systems could be manufactured. In addition, the control of molecular weight range and the ability to vary the back-bone of the polyester permitted the polyester prepolymer manufacturer to drastically vary the physical characteristics of the cured urethane elastoplastic. In this regard, the polyester manufacturers were able to develop a wider range of physical characteristics and considerably more "dynamic" polymers than were those who pursued the PTMG polyether systems. Polyester technology has become so sophisticated that it is possible to manufacture unit polyester polymers of varying molecular weights – blend these to attain a variation of physical characteristics. This approach – that of reacting a mol of isocyanate with a unit polyester – endcaps the polyester system with reactive isocyanate, thereby eliminating the potential hazard of the TDI vapor. In addition, this system improves the dynamic characteristics of the cured polymer system, eliminating the TDI biuret hard segment of the polymer which can occur when free unreactive isocyanate is included in the prepolymer mix.

IMPROVEMENT OF POLYESTER PREPOLYMERS

If a comparison of the polyester/TDI prepolymer of ten years ago is made vs. a like system today, very little difference will be noted in the static physicals of the cured elastomer. However, if the qualities of the prepolymer itself, i.e., in heat stability and the viscosity at mixing temperature considered, today's resins will be considerably less viscous and will have stability built into them so that they may be used through 24 hour heat cycles.

The past decade has resulted in the following improvements in polyester prepolymer cured elastomer systems:

1. Significantly lower cost.
2. Improved low temperature properties.
3. Dramatically improved production casting capability.
4. Dramatically improved product uniformity.
5. Dramatically improved prepolymer stability.
6. Dramatically improved cured polymer dynamic qualities.

This very substantial gain in qualities of the polyester prepolymer systems has resulted in the market currently being primarily oriented to the polyester, whereas in 1960 the cast systems were predominantly polyether.

As was mentioned earlier, the PTMG polyether systems were really the basis for establishing the large scale casting qualities of the urethane prepolymer resin systems in this country. The current market, although ester oriented, still utilizes 40 to 50% of the PTMG polyethers in the high quality elastomers and if the one-shot systems are considered, this ratio may even be higher. To date there are only two major chemical companies marketing PTMG/TDI prepolymer systems on a broad scale, and only one company marketing the PTMG/MDI systems.

In the PTMG prepolymer systems, the variation in the physical characteristics can be derived by the use of either TDI or MDI as the chain terminator and subsequent coreactant – with a curative system. Just as in polyester technology, the molecular weight of the PTMG is important in deriving the cured physical characteristics of the polymer as are the individual isocyanates used. The MDI prepolymer system results in cured elastomers with better oil and chemical resistance and higher modulus characteristics. Even though these systems are higher modulus, they generally will have better flex life and better resistance to propagation of tear in an initiated flexing environment. Also as in polyester technology, the MDI based PTMG prepolymers permit a wide variation in the physical characteristics of the cured polymer depending upon the curative or coreactant used with the MDI/PTMG prepolymer. In the TDI prepolymer systems, varying molecular weight prepolymers are generally designed to be reacted with a primary aromatic diamine. The variation in the prepolymer molecular weight results in the variation in properties.

DEVELOPMENT OF TESTS AND STANDARDS

In the past two years the significance of the PTMG prepolymer systems has come to the forefront with the findings of Gahimer, et al, in the work done to attempt to understand the failure of polyurethane potting systems under high humidity conditions. In these studies, it was found that the normal commercial mode of testing, i.e., 100% R.H., was considerably less arduous environment than conditions wherein the relative humidity is maintained at just below the dew point. For example, at 95% R.H. and 97°C., the half life of some well known polyester systems was less than 4 days vs. 28 days at the same temperature and 100% R.H.

Based upon the work of Gahimer, it appears that the PTMG–MDI–OH cured system has an extrapolated life expectancy in excess of 25 years in high humidity environment such as would be encountered in the tropics or jungle areas.

Gahimer's work was initiated because of problems that were encountered in the stability of potting components in the NAFI warehouse in Indianapolis. The failures encountered in Indianapolis caused each of the branches of the Military to reinvestigate previous anomalous failures encountered on location all over the world. Word of the "hydrolytic instability" of the urethane polymers traveled far and wide very quickly, resulting in similar searches of failures that had been encountered in such diverse locations as steel mills, printing roll houses, automotive, etc. Each of the manufacturers of the articles as well as the end users asked the questions of the prepolymer suppliers as well as initiating test programs on their own. From this very extensive work, the following conclusion can be drawn: Accelerated testing at high temperature may be used to find the resistance to polyurethane and other polymers to normal environment high humidity conditions that may be encountered on earth. The standard test procedures appear to include a 28 day exposure at 97 to 100°C. and 95% R.H. maximum with care being taken to prevent heat radiation effects which would cause precipitation of the moisture onto the test samples. In addition, it has been found that microbiological attack may also adversely affect this system, resulting in failures of the prepolymer systems, therefore, in such standards as MIL-C-5272 or MIL-C-810- the standard Milirary fungus exposure test—rather than merely examination of the specimens after exposure tc see if fungus growth has occurred, samples are actually being tested for the per cent retention of their physical characteristics.

The accumulation of in-field data as well as test data accumulated in the above tests indicate that the total resistance to arduous high humidity environment is best served with the PTMG polyether systems. Of these, the polyether – MDI – OH cure is most resistant, followed next with the PTMG – TDI – aromatic diamine cure.

The polyethylene glycol and polypropylene glycol ether systems may be resistant but the combination of the coreactants with urethane polymers based on these back-bones, is important. Therefore, each individual system would have to be checked in total environment.

Polyester systems have been developed which include stabilizers and fungustats that appear to withstand exposure to high humidity environments. However, these same systems when subjected to high humidity – high temperature (below the dew point) and fungus environment have generally failed catastrophically. It thus appears that the PTMG back-bone based polyurethane elastomers seem to be inherently resistant to the effects of high humidity, fungus, and based on some work done by the Navy, to marine growth. To date, the PTMG based polyurethanes appear to be considerably more resistant to exposure than many other conventional elastomers. It further appears that the work of the Military will result in the issuance of a number of specifications which will specify the poly tetramethyleneether glycol. The work that is currently under way by a number of the automotive companies and others also will probably result in a like specification. Therefore, it appears that the PTMG polyether systems may well attain an unchallenged position in uses wherein long term durability under normal environment is considered.

CAPROLACTONE POLYMERS

A new polymer intermediate is currently being introduced to the market place within the U.S., i.e., the caprolactone systems. Polyurethane prepolymers based on this raw material series have been in use in Europe for a number of years but have not caught on in the U.S. The caprolactones appear to be affected by the type isocyanate used, the type curative used, in much the same manner as has been previously found with both the polyether and polyester systems. The caprolactones' forte seems to be that they are a combination of both polyester and polyether, and the properties of the cured prepolymer reflecting this combination of properties. The caprolactone based polymers tend to have better hydrolytic stability characteristics than the polyesters, yet they have many of the toughness and oil and solvent resistance characteristics normally associated with the polyester systems. As far as is known today, there is only one major supplier of caprolactone based urethane prepolymer systems, although several of the major suppliers of prepolymers are considering these materials.

"B" stage resin systems also are becoming a factor in the industry. These systems are based upon the use of either the prepolymer or one-shot technology wherein the coreactants of each of the systems are mixed and blended as the normal method, but then cooled quickly to stop the reaction.

Generally, in production this is accomplished by utilizing the conventional mixing machine casting the polymer mix onto a belt which is sufficiently long to permit the polymer to react sufficiently to attain intermediate physical characteristics. The slab which is removed from the belt is then granulated and stored at room temperature or lower under anhydrous conditions until used. These granulated resins may then be molded utilizing screw injection molding machines, transfer molding, ram injection molding presses, and conventional compression molding equipment.

The chopped granule of the "B" stage resin is fed into the molding machine, then injected into the preheated cavity (normally set at normal curing temperatures and then held in this configuration until cured, exactly the same manner as would be more normally purchased thermosetting resins).

The "B" stage technique often is used by the larger molder who manufactures his own polymer rather than purchasing it from the outside. This gives the molder obvious advantages in that he is able to formulate the polymer to his immediate needs or technical requirements for the finished part.

Compression molding of the liquid polymer is a technique that is oftentimes used for small production runs of small parts. In this technique prepolymer is normally used as a polymer intermediate. The prepolymer is heated to normal mixing temperatures via a batch system and the curative added. The system is mixed up and poured or cast into the open cavity of the compression mold. The polymer is then allowed to gel to the point where sufficient polymer integrity is developed so that on closing the mold, entrained air will be squeezed from the part. After the appropriate cure times, the part will be totally cured and have the advantage of being molded to size in virtually any configuration.

Open casting still accounts for the major portion of the urethane polymers used in this country. In casting, the prepolymer and curative or coreactants are heated to the appropriate temperatures in mixing machines, then mixed, and "poured" from the dispensing head of the mixing machine into an open cavity. There are many techniques involved in open casting which have developed through the years to eliminate entrapped air, flow lines, etc. Casting has many advantages, particularly in the manufacture of large parts wherein it is possible to cast the parts to dimensions required with a minimum of effort and a minimum of mold cost.

One-shot technology, although in use to a major degree in Europe, has not really grown in this country. Originally, most of the low durometer print rolls and other low modulus applications in this country utilized the one-shot technology, i.e., the polyol plus varying degrees of isocyanate which will react to form the cured polymer. This particular system may utilize either polyethers or polyesters in the polyol portion and any one of the isocyanates to coreact to form the cured polymer. In this instance polymeric isocyanates are often used to develop specific properties as are di- and trifunctional polyols. In some instances aromatic diamines are also included in the system, not only to serve as coreactants, but also as catalysts.

More recently low modulus prepolymer systems appear to be replacing the one-shot systems in production. Even though the prepolymer route will increase the overall cost of the system (must include additional reaction), the prepolymer route appears to yield far simpler production characteristics, resulting in lower scrap rates, more uniform cures, and a higher degree of polymer properties in the cured polymer. For example, in print roll technology the use of Uniroyal Vibrathane 6005 with various curatives has become a major factor in the industry due to the durability of the finished roll in service as well as the ease of production to develop "on spec" finished parts.

During the past decade polyurethane elastomers have become a commercial item. Many of the problems that were inherent with the polymers that were available ten years ago have been eliminated. Many new production techniques have been developed to enable the parts manufacturer to manufacture finished parts with good integrity and engineering qualities on a routine production basis.

The urethane elastomer market will continue to grow as the advantages of durability and outstanding performance of these polymers becomes known through industry. The close cooperation between the supplier of resins, the manufacturer of the parts, and the design engineer is continuing. This effort will result in major growth of the market so that by the end of this decade polyurethane elastomers will probably have attained the status of a "commodity."

Table 1. Leading U.S. Elastomeric Urethanes Producers*

	Capacity (million lbs./year)			
Producer and Location	**Castable**	**Millable**	**Injection Moldable**	**Others**
American Cyanamid Bound Brook, New Jersey	5	1	–	Spandex
DuPont Wilmington, Delaware	30	1	–	Corfam Spandex
B. F. Goodrich Chem. Avon, Ohio	–	–	15	Aztran
Goodyear Tire & Rubber Akron, Ohio	15	–	–	–
Mobay Chemical Co. Pittsburgh	10	–	10	–
Thiokol Chemical Trenton, New Jersey	3	1	–	Adhesive
Uniroyal, Inc. Naugatuck, Connecticut	5	1	1	Spandex
Witco Chemical Chicago	8	–	–	–

*In all, there are about 85 producers with 50,000 to 1 million lbs./year capacity.

Table 2. Urethane End Uses 1969 U.S. Consumption, Million Pounds

Castable	
Fibers	12.0
Synthetic Leather	6.0
Industrial Tires	5.0
Rolls	5.0
Proofed Goods	2.5
Other	7.5
Injection Moldable	16.0
Total	54.0

REFERENCES

1. C. F. Blaich, "How to Cast Urethanes.", *Plastics Technology,* November, 1968.
2. C. F. Blaich and A. J. Sampson, "The Effect of Isocyanate Structure on Cured Urethane Elastomer Properties." *Chemical and Engineering News,* September 26, 1960.
3. C. F. Blaich, "The Structure-Property Relationship of a New Series of Polyurethane Elastomers.", *Journal of Elastoplastics,* January 1, 1969.
4. J. M. Buist and H. Gudgeon, "Advances in Polyurethane Technology.", Maclaren and Sons, Ltd.
5. R. J. Ferrari, "Polyether and Polyester Urethane Elastomers.", *Rubber Age,* February, 1967.
6. P. A. Gianatasio, "Polyurethane Polymers.", *Rubber Age,* July and August, 1969.
7. A. J. Sampson and C. F. Blaich, "Effect of Amine Structure on the Properties of Cured Polyurethane.", *Rubber Age, 89,* May, 1961.
8. J. H. Saunders and K. C. Frisch, "Polyurethane Chemistry and Technology, I and II.", INTER-SCIENCE.
9. T. M. Vial and C. F. Blaich, "Faster Methods of Physical Testing.", *Rubber Age,* May, 1962.
10. C. Magnus, R. A. Dunleavy and F. E. Critchfield, "Stability of Urethane Elastomers in Water, Dry Air and Moist Air Environments.", *Rubber Chemistry and Technology,* September, 1966.
11. F. H. Gahimer and Fred W. Nieske, "Hydrolytic Stability of Urethane and Polyacrylate Elastomers in Humid Environments.", *Journal of Elastoplastics,* October, 1969.
12. F. H. Gahimer and F. W. Nieske, "Navy Investigates Reversion Phenomena of Two Elastomers.", *Insulation,* August, 1968.

CHARLES F. BLAICH, *Uniroyal Chemical Company*

CHAPTER 7

MICROCELLULAR URETHANES

INTRODUCTION

Flexible urethane polymers, as low density foams and solid elastomers, are well known materials. Low density urethane foams are extensively used for comfort cushioning in upholstered furniture and automotive seating. In these applications the plush feel, high resilience, and good shape retention properties of the material have led to rapid consumer acceptance. Similarly, flexible solid elastomers with excellent abrasion and solvent resistance, high tear and tensile strength, and good elongation are used in industrial applications where outstanding performance is required.

Coming as a relatively new development, microcellular urethanes combine many of the high strength features of solid elastomers with the unique cushioning and load bearing of high density flexible cellular structures. These microcellular foams have been adopted for a growing number of diverse products. Such items as auxiliary springs, truck dock pads, shoe soles, gaskets, vibration isolators (1-3), and more recently exterior automotive bumpers (4-6) have been fabricated using microcellular foams.

This paper will review the chemistry, properties, processing, and applications of microcellular urethanes. Special emphasis will be given to the use of this foam in some automotive bumper production.

MICROCELLULAR FOAM CHARACTERISTICS

Microcellular urethanes have been termed expanded elastomers and also high density foams. Both definitions apply since the polymer chemistry is essentially that of urethane elastomers, and the most useful commercial forms have densities ranging from 20 to 50 lb./cu. ft. In most cases the foam has a characteristically fine cell structure, hence the term "microcellular" has been adopted.

For a proper perspective on the general properties of microcellular foams, Table 1 cites a comparison with low-density flexible foam, polyether urethane elastomer, and polyester urethane elastomer. Values shown are representative data for the general types and it should be understood that within a particular category, materials with a fairly broad properties range may be compounded.

In addition to good high strength properties, microcellular foams can be compressed with very little sideway bulge (2). Figure 1 illustrates this behavior in a comparison with solid rubber. This property can be important where installations of shock mounts or vibration dampeners are made in confined places.

Depending on the compounding of microcellular foams, either open-celled

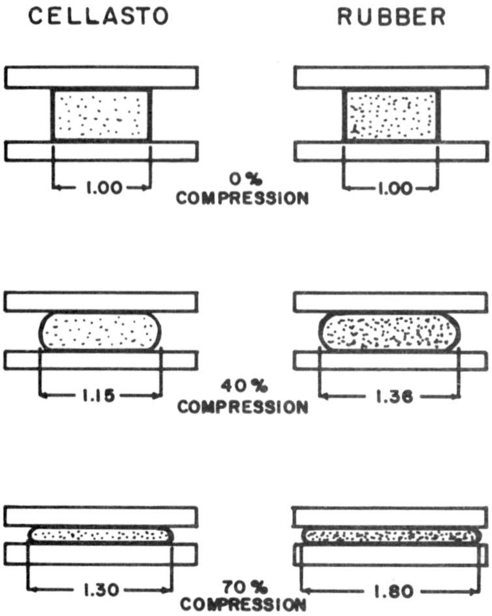

Figure 1. Compressibility of Microcellular Foam Without Sideway Bulge is Shown in a Comparison to Solid Rubber.[2]

Table 1. Urethane Polymer Properties.

	Microcellular Foam (7)	Low Density Cushioning Foam (8)	Polyether Elastomer (9)	Polyester Elastomer (10)
Density,	45	1.8	70	70
Tensile strength, psi	650	20	4600	7000
Elongation, %	170	260	840	700
Tear strength, lb./in.	150	3	210	450
Compression load, psi	750	1.2	–	–
Hardness, Shore A	65	–	75	85

(a) 50% Compression
(b) 65% Compression

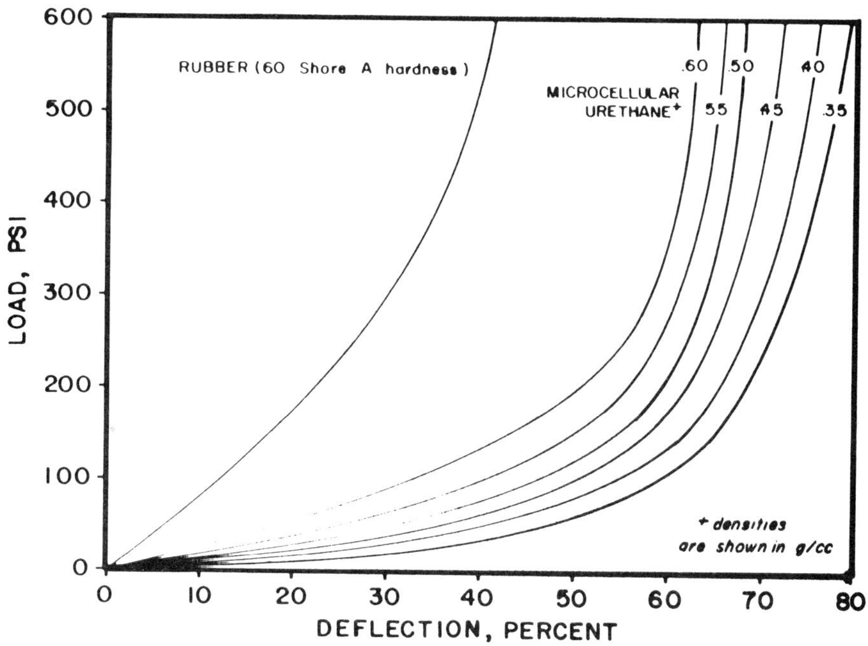

Figure 2. Compression Load-Deflection Behavior of Microcellular Foam. Densities Shown Range From 22-38 lbs./cu.ft.[2]

or predominantly closed-celled foams can be made. For end uses such as shoe soles, low water absorption is important and here closed-celled foam is preferred. In one study it was shown that no noticeable difference in water pick-up occurred after 200 flexes to 50 per cent deflection with closed-celled foam (3). However, closed-celled materials generally have higher compression set values than open-celled foams of the same density.

Microcellular foams have excellent load bearing capacity and fatigue resistance. These properties are especially important for such applications as motor mounts where heavy loads must be maintained in original positions for extended periods. For example, a closed-celled SBR compound of 35 lb./cu. ft. density at 10 per cent deflection required a 30 psi load, while for microcellular foam of the same density the required load was 220 psi (3).

A related performance characteristic is the behavior of microcellular foam under dynamic loading. Burck has shown that under increasing load, the material becomes stiffer because of the mass effect becoming operative (2) (11). Figure 2 illustrates how microcellular urethane foam differs from conventional rubber under increasing load. The stiffening of the microcellular foams with greater applied loads provides a self-compensating cushioning material that responds gently to light loads but becomes firm under heavy loads. Further, as the velocity of impact increases, the load deflection curve is steeper. This results in a "soft" response at low-velocity impacts, while

Table 2. Microcellular Foam Ingredients.

Polyol
Aromatic diamine
Surfactant
Catalyst
Blowing Agent
Isocyanate

with high-velocity impacts a stiffer response is obtained (11).

Along with strength properties approaching that of elastomers, but with a reduction in density, and unique cushioning features, microcellular foams can be molded in large masses. Unusual contours can be reproduced with molded parts that exhibit insignificant shrinkage and smooth surfaces. The moldability of this material was of paramount importance in the development of microcellular urethane auto bumpers.

Finally, these foams show excellent recovery from deformation. By formulation tailoring, materials with high resiliency or with low rebound characteristics can be made. Again, the good recovery feature was essential in the exterior automobile bumper development.

CHEMISTRY OF MICROCELLULAR FOAMS

The final properties of microcellular foams can be closely controlled by polymer design. Formulation technology has been extensively examined by a number of investigators (3) (7) (12) (13) (14). These foams may be made by one-shot and prepolymer methods, using tolylene diisocyanate (TDI) or polymeric isocyanates. Expansion can be effected by using water, monofluorotrichloromethane, methylene chloride, and chemical blowing agents that decompose to release nitrogen. Processing, reactivity and final properties are influenced by choice of polyol(s), reactive amine, and the polyol/amine ratio. Surfactants are used to control the open or closed-cell structure while the type and concentration of catalyst provide control of reaction rate and molding characteristics. These variables, and others, allow the urethane technologist wide compounding latitude in formulating systems to achieve specific end properties.

Table 2, then, cites a generalized listing of ingredients for microcellular foam. The polyol/amine/isocyanate choices, and concentration primarily determine polymer backbone structure and therefore, finished polymer properties. A reactive aromatic diamine is incorporated in most systems to improve strength properties. Surfactants used are the dimethyl silicone oils when predominantly open-celled structures are desired, and dimethylsiloxane-polyglycol copolymers where closed-celled foams are required. A variety of tertiary amine and organo-metallic compounds have been reported as useful catalysts. Blowing agent concentration is the controlling factor in foam density which in turn has a profound effect on other physical properties.

NON-REACTIVE BLOWING AGENTS

In some of the earlier work on blown elastomers, prepolymers based on TDI and polytetramethyleneether glycol, and Nitrosan (duPont's nitrogen-releasing blowing agent with N,N'-dimethyl-N,N'-dinitroso terephthalamide as the active ingredient) as the blowing agent were used to prepare a variety of foams (12) (15). Table 3 cites the formulations used and properties achieved with this approach. In this work, the diamine used was methylene bis (o-chloroaniline). Ingredients were cast into 212°F molds and cured at the same temperature to bring about polymer expansion and cure. It is also interesting to note that when a blend of a low molecular weight diol and a triol was used as the curing mixture, strength and load-bearing properties were dramatically lowered.

In other investigations (3) (16), monofluorotrichloromethane and methylene chloride have been used as expansion agents. As shown in Table 4, foams were

Table 3. Chemically Blown Prepolymer Foams (12).

Ingredients	1	2	3
Adiprene L-100 (Dupont)	100	-	100
Adiprene L-167 (Dupont)	—	100	—
Cab-O-Sil M-5 (Cabot)	5	5	3
DC-200 (50 ctks) (Dow Corning)	1	1	1
1,4 Butane diol	—	—	3.2
Trimethylolpropane	—	—	0.9
MOCA[a]	12	20	—
Stannous octoate	—	—	0.5
50% Nitrosan[a] dispersion in dioctylphthalate	2	2	4
Properties			
Density, lb./cu. ft.	22	23	15.3
Tensile strength, psi	325	500	54
Elongation, %	325	300	350
Compression set, (50% Comp.), %	33	35	7.5
Compression load, psi, 25%	44	92	4.4
50%	90	175	9.2
Resilience, Yerzley, %	60	35	60.5

Table 4. Solvent Blown Microcellular Foams (16).

Ingredients	1	2
Prepolymer (6.4% -NCO)	100	100
Methylene bis (o-chloroaniline)	19.5	19.5
Adipic Acid	2.5	2.0
Carbon black	—	2.0
Silicone surfactant	—	2.0
Dioctylphthalate	0.5	—
Monofluorotrichloromethane	3.4	—
Methylene chloride	—	6.2
Properties		
Density, lb./cu. ft.	20	16
Water pick-up, Vol. %	9.7	2.3

prepared by reacting a low free -NCO prepolymer (based on the reaction of polytetramethyleneether glycol and TDI) and methylene bis(o-chloroaniline) as the curative. In these foams the reaction exotherm volatilized the inert low-boiling solvents to expand the polymerizing mass. In addition to methylene bis(o-chloroaniline), a mixture of cumene diamine/m-phenylene diamine, 2,4-diamino-toluene, and a mixture of m-phenylene diamine/methylene bis(o-chloroaniline) were reported used as curing agents. The systems were characterized by rapid reactivity and short tack free times.

WATER BLOWN FOAMS

When water is included in a microcellular foam formulation, the isocyanate-water reaction results in carbon dioxide as a reaction product. This evolution of carbon dioxide gas, which is trapped in the polymerizing mass, acts as the expansion agent. Final foam density is, of course, dependent on the initial water concentration in the foam mix.

Saaty and co-workers used a one-shot technique to make water-blown foams where methylene bis (4-phenyl isocyanate) (MDI) and 3,3′-dichloro-benzidine were reacted with a number of hydroxyl-terminated resins (13). In one series, as shown in Table 5, isocyanate index had a direct influence on many important physical properties. Especially sensitive to this variable were properties such as compressive strength, tensile strength and ultimate elongation.

Foams made in the 40-55 lb./cu. ft. range by Bailey and co-workers showed a density-related spread in physical properties (7). Table 6 cites these properties where the polyol/methylene bis(o-chloroaniline) equivalents ratio was 1/1.

Table 5. One-Shot MDI Microcellular Foam (13).

Basic Formulation			
Isonate 143 L (Upjohn)			Vary
1000 M.W. polyester diol			100
3,3′-dichlorobenzidine			15.1
Silicone surfactant			2.0
Stannous octoate/DOP			0.1
Dabco R-8020 (Houdry)			0.7
Water			0.25
Pigment			2.0
Isocyanate Index	0.95	1.05	1.15
Foam Properties			
Hardness, Shore A	26	40	60
Density, lb./cu. ft.	34.3	30.0	30.2
Tensile strength, psi	200	448	492
Elongation, %	508	210	93
Tear strength, die C, lb./in.	58	121	110
Compressive strength, psi			
25% defl.	28.4	84.4	148
50% defl.	103	187	295

The polyol mixture was a blend of a 1000 M. W. polyether diol and a 3000 M. W. polyether triol.

When the ratio of methylene bis(o-chloroaniline) to polyol was varied in water-blown, TDI-based foams, a steady increase in the load-bearing properties was noted as the concentration of aromatic amine increased (7). This trend and other property changes are shown in Table 7. The increased amounts of amine resulted in considerable polymer stiffening because of the greater concentration of urea groups and aromatic ring structure in the polymer backbone.

Table 6. One-Shot TDI Microcellular Foam (7)
Effect of Density on Foam Properties.

Polyether Av. Functionality	=	2.2
Polyether Av. Molecular Wt.	=	1400
TDI Index	=	1.05

Density, lb./cu. ft.	43	48	50
Tensile strength, psi	620	740	870
Elongation, %	190	180	200
Hardness, Shore A	65	69	76
Compression load, psi, 50%	760	980	1600

Table 7. One-Shot TDI Microcellular Foam (7)
Effect of NH_2/OH Ratio on Foam Properties.

Polyether Av. Functionality	=	2.0
Polyether Av. Molecular Wt.	=	1020
TDI Index	=	1.1

NH_2/OH equivalents ratio	0.75	1.0	2.0	2.5
Density, lb./cu. ft.	41	42	50	53
Tensile strength, psi	575	580	1110	1635
Elongation, %	190	215	165	90
Hardness, Shore A	61	73	83	93
Compression load, psi, 50%	430	700	1200	2310
Resilience, (G-H rebound), %	36	34	33	32
Compression set, (50% comp.), %	13	26	18	25

Table 8. One-Shot TDI Microcellular Foam
Effect of Polyol Blend on Foam Properties.

1000 M.W. Diol/4500 M.W. Triol (Eq. Wt.)	7/1	4/1	1/1
Density, lb./cu. ft.	44	40	42
Tensile strength, psi	710	570	420
Elongation, %	100	110	120
Hardness, Shore A	81	71	66
Compression load, psi, 50%	840	620	450
Resilience, (G-H rebound), %	38	42	51
Compression set, (50% comp.), %	10	19	11

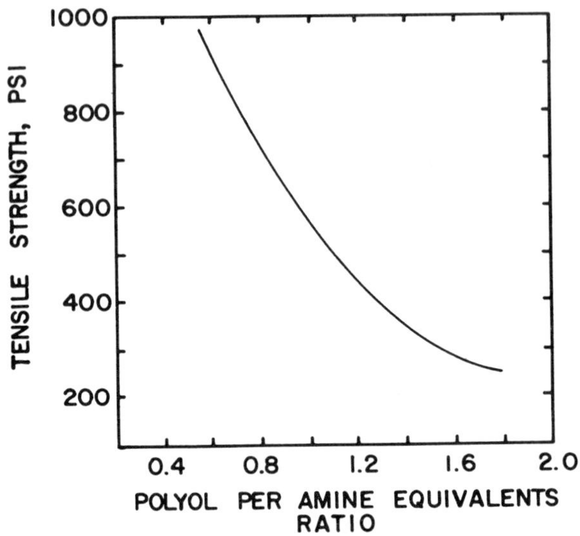

Figure 3. Effect of Amine Content on Tensile Strength of Microcellular Foam.

A similar trend in microcellular foam properties was observed when the OH/NH_2 ratio was varied, using a liquid aromatic amine in the formulation (14). In this study the experimental aromatic amine used was a condensation product having di- and polyfunctionality, liquid characteristics at room temperature, and an equivalent weight of 128. The polyol blend was a 2/1 mixture by weight of a 1000 M. W. secondary hydroxyl-terminated diol and a 3000 M. W. secondary hydroxyl-terminated triol.

Foams made in the 0.8-1.5 OH/NH_2 ratio showed good molding properties. The effects of the formulation changes on foam tensile strength and compression load are illustrated in Figures 3 and 4.

By manipulating the type and ratio of polyether polyols used in one-shot TDI foams, additional property changes can be achieved. For example, Table 8 demonstrates the changes affected in foam properties where the ratio of a 1000 M. W. diol to a 4500 M. W. primary hydroxyl-terminated triol is varied.

PROCESSING MICROCELLULAR FOAMS

In the actual foaming operation to make microcellular urethanes, liquid ingredients are pumped, in the proper ratio, to a high shear dispensing mixer. Where solid diamines are used, these are first dissolved by heating with the polyol.

With water-blown foam, the control of moisture content is critical. For example, for a 40-45 lb./cu. ft. foam, only about 0.2 parts of water are required per hundred parts of polyol. Therefore, precautions must be taken during preblending operations to prevent atmospheric moisture pickup. In some

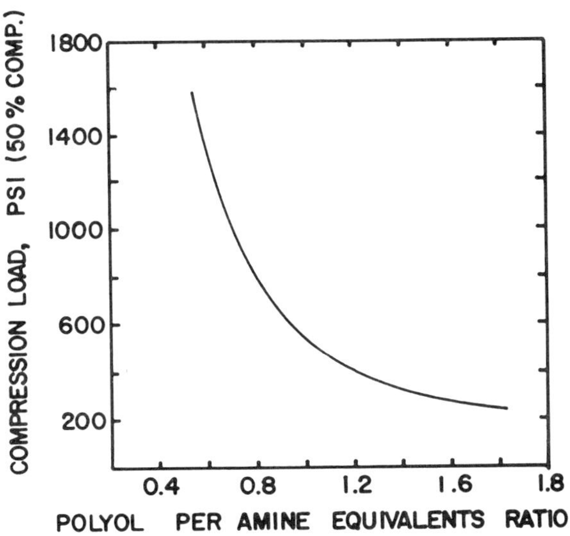

Figure 4. Effect of Amine Content on Load Bearing Properties of Microcellular Foam.

instances, it may be necessary to heat the polyol/amine blend under vacuum to dehydrate the ingredients. When starting new work with freshly compounded materials, it is desirable to make a test foam prior to any addition of water. Depending on the test foam density, water is then added to the polyol premix at the proper level to arrive at the desired density. It has also been shown that molecular sieves are useful additive drying agents where the water concentration in the polyol premix is too high (7).

Because the foam expansion in microcellular urethanes is only about 2 to 3 times the liquid ingredients volume, several factors have an important influence on foam density. These are:

1. Concentration of foam blowing agent
2. Mold temperature
3. Unit mass to be foamed
4. Mold mass and material of construction
5. Ingredient temperature
6. Entrapped air
7. Molding pressure

For consistent reproducibility, all of these elements must be closely controlled.

Where products with smooth surfaces are required, such as auto bumpers, molds must be highly polished. Release agent applied must also be buffed to a high-gloss, mar-free finish prior to casting foam into the mold.

In most production systems, the molds are used hot (150-200°F), and parts are post-cured at elevated temperatures to obtain good physical properties. Some work as shown, however, that by proper catalysis, molding at more moderate temperatures (80-120°F) is possible.

INDUSTRIAL APPLICATIONS

As pointed out earlier in this paper, microcellular foams have special properties that make these materials useful in a number of applications. This section will highlight some of these end uses.

1. Automotive Bumpers (5) (6) (17) (18)

Microcellular foam exterior bumpers are a dramatically new concept in automotive styling. Introduced by Pontiac in 1968 on the GTO model as the "Endura" bumper, these molded components are painted with a flexible urethane coating to match the body color. Able to withstand the low-speed bumps and bruises of parking lot maneuvering without damage, the units are also free of corrosion, and are resistant to stone pitting.

The popularity of the microcellular foam bumpers is indicated by this listing of 1968-1970 uses.

1968 - Pontiac GTO front bumper
1969 - Pontiac GTO front bumper
- Pontiac Bonneville and Catalina front grille divider
- Pontiac Bonnevile and Catalina rear bumper applique (option)
- Chevrolet Camaro bumper (option)

1970 - Pontiac GTO front bumper
- Pontiac Firebird front bumper
- Plymouth Barracuda front bumper (option)

2. Shock Mounts (1) (2) (11)

High-speed racing cars have had microcellular urethane pads installed to cushion the centrifugal force acting on the outboard suspension systems when the vehicles go into sharp curves. Similarly, the foam is used as impact bumpers from such needs as limit stops for delicate instruments, to heavy duty crane stops.

3. Vibration Isolators (1) (2) (11)

Foam pads serve as excellent vibration dampeners, with good fatigue properties for suspension systems of heavy-duty trucks, off-the-road vehicles, and textile machinery.

4. Shoe Soles (3) (19)

Closed-celled blown urethane elastomers give outstanding service as lightweight footwear soles. The low water pickup and excellent flex properties provide long product life.

5. Auxiliary Springs (1) (2) (11)

In Germany these materials are used as safety bumpers in elevator shafts, replacing steel springs. The foams are also used as jounce bumpers on trucks to provide overload protection. In Italy the foam is used in combination with mechanical shock-absorbers in passenger cars.

6. Wheels and Rollers (1) (2) (11)

In sheet-metal stamping plants, rollers of microcellular urethane are used to convey metal panels. Here, the resistance to cuts, tears and lubricants, and the quiet performance of the urethane elements have proven superior to other materials. Other installations include back-up rolls for sanding and abrasive belts, and soft-roll coverings in paper and packaging plants.

7. Seals and Gaskets (1) (2) (11)

It has been reported that microcellular products, such as seals for ball joints in vehicle suspension systems, have given excellent performance. The resistance to abrasion, small side-ways expansion and high compressibility of the foam offer improvements over previously used materials.

Other applications include orthopedic devices, recoil absorbers, and spring elements for railroad cars. Additional uses will develop as design engineers become more familiar with the properties of microcellular urethanes.

SUMMARY

Microcellular foams have special properties which approach in strength those of urethane elastomers, while showing the compressibility of flexible cellular materials. This paper has reviewed these unique properties, the chemistry and processing of the foams, and finally the industrial applications for expanded elastomers.

REFERENCES

1. A. H. Landrock, "Polyurethane Foams: Technology, Properties and Applications", Plastec Report 37, Plastics Technical Evaluation Center, Picatinny Arsenal, Dover, N. J. , (1969), pp 50-51.
2. R. C. Burck, "Microcellular Urethane Elastomers Breakthrough in '68", 2nd SPI International Cellular Plastics Conference, New York, New York, Nov. 1968.
3. D. Bianca and R. E. Knox, *Rubber Age,* V 98, May (1966), pp 78-80.
4. *Motor Trend,* Feb. (1968).
5. *Modern Plastics,* Oct. (1967), pp 89-92.
6. R. W. Malcolmson, Rubber Age, V 101, N 10, (1969), pp 74-77.
7. M. E. Bailey, M. Kaplan, B. Taub and G. Wooster, *J. of Cell. Plast.,* V 5, N 4, (1969), pp 237-246.
8. J. H. Saunders and K. C. Frisch, "Polyurethanes, Chemistry and Technology, Part II Technology", Interscience Publishers, New York, New York, (1964), p 72.
9. S. L. Axelrood and K. C. Frisch, *Rubber Age,* V 88, N 3, Dec. (1960), pp 465-471.
10. Witco Chemical Co., Technical Service Bulletin No. F-13, Apr. (1964).
11. Product Engineering, Feb. 24, 1969, pp 61-62.
12. R. O. Menard, "Microcellular Adiprene L Vulcanizates Blown with Nitrosan",

Adiprene Urethane Rubber Bulletin, E. I. du Pont de Nemours & Co., Inc., Jan. 21, 1964.

13. N. N. Saaty, et. al., *J. of Elastoplast.* V 1, July (1969), pp 170-184.
14. L. M. Zwolinski, M. Kaplan and M. E. Bailey, "Molding Castable Urethane Polymers", SPI Cellular Plastics Division, 14th Annual Technical Conference, Detroit, Mich., Feb. 24-26, 1970.
15. J. H. Saunders and K. C. Frisch, "Polyurethanes, Chemistry and Technology, Part II Technology", Interscience Publishers, New York, New York, (1964), pp 111-116.
16. British patent 1,037,907, Aug. 3, 1966.
17. *Soc. of Auto. Eng. Journal,* Dec. 1968, p 106.
18. R. H. Eshelman, *Rubber Age,* V 100, N 9, (1968), pp 68-73.
19. K. Murai and K. Fukuda, *J. of Elastoplast.,* V 1, July 1969, pp 150-160.

L. M. ZWOLINSKI, *Specialty Chemicals Division, Allied Chemical Corporation*

CHAPTER 8

URETHANE COATINGS FOR MICROCELLULAR BUMPERS

INTRODUCTION

The rapid growth and wide acceptance of urethane polymers over the last decade can be attributed to their unusual combination of toughness, abrasion-resistance, low temperature elasticity and durability. In addition, a wide variety of formulating parameters offers unlimited versatility in the application requirements of foams, coatings and elastomers.

More recently high density flexible urethane foams, generally referred to as foamed urethane elastomers or microcellular urethane foam, have been successfully used in the automotive industry for bumpers [1].

The physical properties of these high density flexible urethane foams can be varied over a wide range by changes in formulation or mechanical processing. Molded parts have a tough continuous skin which reproduce the exact surface and contours of the mold. Cast microcellular urethane is produced by mixing and reacting two liquid components in a high shear dispensing mixer. The resulting liquid mixture is deposited into a highly polished mold. Expansion of the reaction mass forces the blown elastomer to fill the mold cavity and to encapsulate a metal bumper mounting insert. After a suitable cure, the bumper is removed from the mold. The microcellular urethane foam has an average thickness of about one quarter of an inch over the steel member. Ward [2]estimates front bumper usage for cast elastomer in 1970 at 210,000 units with an average weight of 13.8 lbs. per car.

Although most automotive bumper applications utilize a microcellular elastomer because of its lower cost on a volume basis, in some instances injection-molded solid urethane elastomers have been utilized.

A more recent development is the application of a solid urethane elastomer by means of a two-component spray gun. In present experimental work, the bumper coating is sprayed on in four or more separate applications up to a 15 mil thickness.

In all instances, the success of the expanding usage of urethane micro-cellular foam or solid elastomers for automobile bumpers is dependent upon a compatible, durable, flexible non-discoloring coating, which can be color-matched to the exterior automobile finish.

A variety of color-stable urethane coatings has found applicability in finished automotive exterior parts. Two basic types are: (1) Urethane lacquer and (2) urethane enamel.

URETHANE LACQUERS

Non-discoloring urethane lacquers have successfully been used for exterior microcellular urethane bumpers.

The characteristics of a urethane lacquer system developed by Allied Chemical Corporation [3] are:

-One-package lacquer.
-Excellent color stability.
-Color match to metallic and non-metallic automotive paints.
-High durability and weatherability.
-Outstanding abrasion resistance.
-High tensile strength and elongation.
-Low temperature flexibility.
-High gloss.
-Polishable with cleaners and waxes.
-Resistant to gasoline and oils.

The primary advantage of the one-package urethane lacquer is that drying or curing occurs by solvent evaporation. The final dry film can be achieved by air-drying, moderate heat-drying (ca. 160°F.) or a combination of the two. Drying of the film depends upon the volatility of the solvent system and the number of coatings required. To achieve high gloss, a clear coat of lacquer is applied over the color-matched pigmented lacquer. Since these lacquers do not contain a curative, catalyst or reactive groups, they possess an indefinite pot life. As a result, the urethane lacquer has considerable versatility and convenience in application with existing or available equipment.

Urethane Lacquer Solution Characteristics

Typical solution characteristics for the lacquers are 15,000 cps. (@ 25°C.) at 35% N. V. A wide variety of solvents and solvent blends can be used to reduce solids content suitable for spray application. Xylene, methyl ethyl ketone or isopropanol can be used to some extent as a diluent. For larger dilutions, stronger solvents such as Nadone, dimethylacetamide and methylcellosolve are necessary to avoid precipitation. Typical solvent compatibilities over a range of dilutions with various solvents are shown in Figure 1.

The application and performance properties of the lacquer coating depends on the volatility characteristics of the solvent system and has a pronounced effect on properties such as [4]:

1. Sprayability (degree of atomization, wet or dry spray)
2. Flow or film smoothness
3. Tendency to bubble or pinhole
4. Dry time
5. Blush-resistance

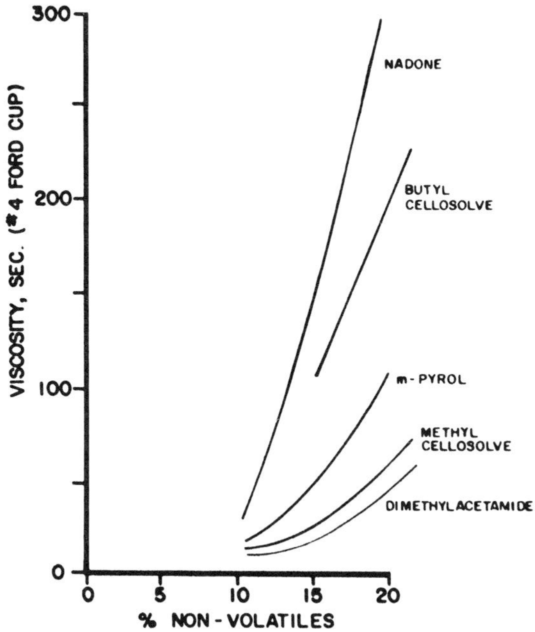

Figure 1. Typical Solvent Compatability for new Urethane Lacquer.

The rate at which evaporation occurs must therefore be closely controlled by careful selection and blending of solvents and diluents.

A typical solvent blend suitable for general spray application at 10% N. V. is:

Methylethyl ketone	30%
Nadone	30%
Xylene	30%
Butyl Cellosolve	10%

Optimization of the solvent reduction system will be a function of the substrate, spray conditions and available equipment.

Stress/Strain Characteristics of Flexible Films

Figure 2 illustrates typical stress-strain curves for the urethane lacquer film compared with a variety of flexible firms [5]. The urethane lacquer films are significantly outstanding in that they exhibit a true yield point followed by deformation until break. The location of the definite yield point is of

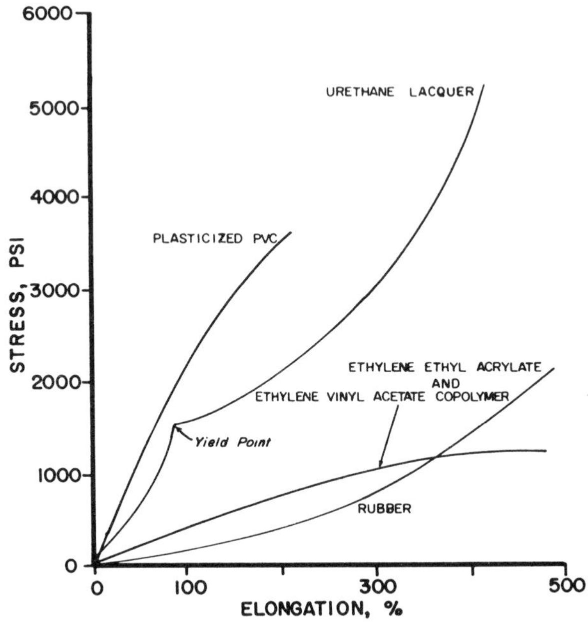

Figure 2. Stress-strain Characteristics for Flexible Films.

considerable importance since its location on the stress axis is an indication of the point of film deformation beyond a region of very rapid recovery. The yield point on the strain axis is a measure of how far the material can be distorted before there is a change in the speed and extent of recovery. In the case of the urethane lacquer films, although high elongations are attained (ca. 350%), a major portion of the work required (distance times force) to deform the sample is encountered within the first 100% elongation period. In contrast with a typical rubber, relatively low resistance is observed at 100% elongation, giving the new system the feel and appearance of enamels, yet physical characteristics of an elastomer.

Durability By Stress/Strain Measurements

Many methods are used to measure the application and performance properties of experimental coatings or to compare one coating with another.

A technique we have used to predict the durability and performance of coating materials is to measure stress-strain characteristics and the tensile strength property change over a range of exposure times [6].

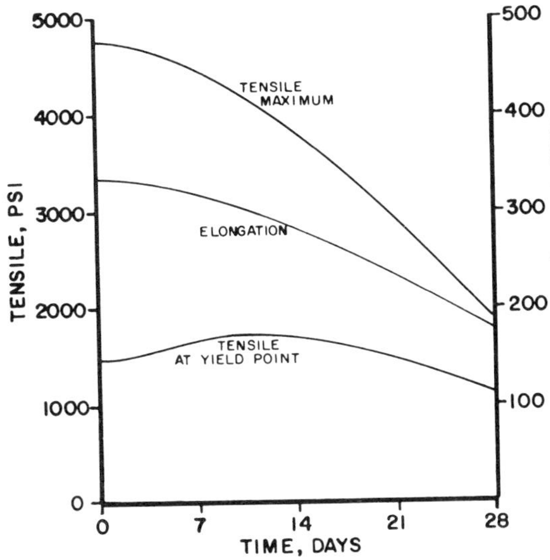

Figure 3. Properties of Urethane Lacquer after Exposure in Hot Water (60°C).

Free films are prepared by drawdowns on photographic ferrotype plates coated with a fluorocarbon release agent. Drying is accomplished at 85°C for 1/2 - 1 hour depending on the solvents present. The dry film thickness is controlled by using an adjustable doctor blade for the drawdown. Table 1 shows the pertinent information on conditions for testing the film on an Instron machine.

Instron data allow the calculation of tensile yield (psi) and tensile maximum (psi) at break. The % elongation at break is obtained from the Instron gauge readings.

By using the techniques described above, durability characteristics for the urethane lacquer film were studied. Table 2 shows typical data obtained after exposure of a urethane lacquer film in a Fadeometer (type FDA-R) at 165°F black plate temperature. The data indicate that after 1,000 hours exposure in the Fadeometer, clear urethane lacquer films retain 70% of their original tensile strength and 50% of their elongations. Table 3 shows the typical data obtained after exposure in a Weatherometer (WXR 150°F black plate temperature) for films made from the new urethane lacquer.

The properties of clear films of the new urethane lacquer after exposure to hot water (140°F) shown in Figure 3 indicate good retention of properties, particularly with respect to film elongation.

Retention of properties of clear films after exposure to elevated temperatures

Table 1. Instron Testing Conditions for Free Films.

Specimen Size:	3½ inch long dog-bone style cut with a die, based on ASTM D-412-51T dimensions. Conditioned 24 hours at 72° ± 2°F @ 50 ± 2% R.H.
Specimen thickness:	3.5-4.0 mils.
Number of Specimens:	8
Jaw Spacing:	1½ inches, pneumatic jaws with neoprene jaw face.
Cross head:	20 inches per minute.
Full-Scale Setting:	10-20 pounds.
Chart Speed:	10 inches per minute.

Table 2. New Urethane Lacquer. Typical Fadeometer Exposure (Type FDA-R) of a Free Film

	Tensile Strength (psi)		
Exposure, Hours	Yield	Maximum	% Elongation
0	1600	5000	250
1000	2360	3500	125

Table 3. New Urethane Lacquer. Typical Weatherometer Exoosure (Type WXR) of a Free Film.

	Tensile Strength (psi)		
Exposure, Hours	Yield	Maximum	% Elongation
0	1600	5000	250
100	1880	4500	210
200	2380	4500	180
400	2100	3200	140
600	2100	2500	72

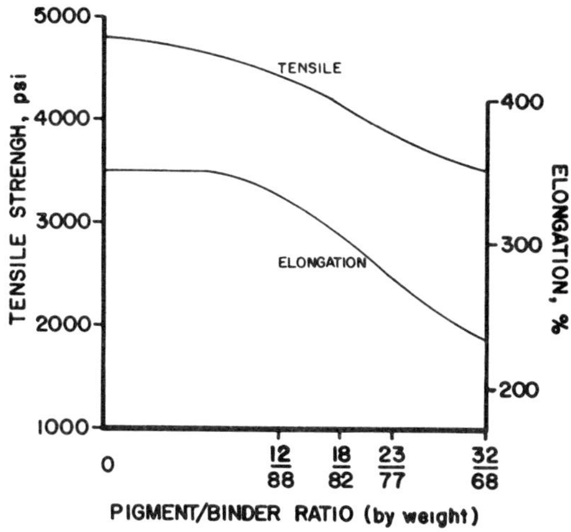

Figure 4. Tensile Strength and Elongation Properties of Pigmented Urethane Lacquer at Various Pigment-Binder Ratios Rutile Titanium Dioxide R-900.

is shown in Table 4. The results indicate that at temperatures up to 338°F, little or no change in tensile strength or elongation is observed. Other typical film properties for the urethane lacquer are shown in Table 5.

Various types and concentrations of pigments and fillers exhibit a marked effect on the physical properties of a paint film. The influence of pigment volume concentration (PVC) or pigment binder ratio (P/B) has been studied for the urethane lacquer with respect to free film strength characteristics. Figure 4 shows the effect of Rutile Titanium Dioxide at various concentrations on tensile strength and elongation of the film.

Pigment Dispersion

Pigments are generally dispersed or ground into the urethane lacquer in various ways. In the grinding process, pigment aggregates are broken up and the dispersed particles are wetted by the urethane resin. Conventional grinding techniques using a pebble mill or sand mill have successfully been used with the urethane lacquer. A wide variety of organic pigments generally suitable for use with the urethane lacquers is available from Harmon Colors, Specialty Chemicals Division of Allied Chemical Corporation. These include:

Harmon's Indofast Yellow Toner Y-5 43
Harmon's Quindo Red R-6730
Harmon's Indofast Blue Toner B-79
Harmon's Phthalo Blue B-4714

Harmon's Phthalo Green G-5056
Harmon's Phthalo Green Extra Yellow G-5400

Inorganic pigments such as:

Molybdate Orange
Cadmium Red, Yellow and Orange
Red and Yellow Oxides
Aluminum pigments
Titanium Dioxide

can be used both alone and in combination. Virtually any desired color and shading can be attained with metallic and non-metallic automotive colors.

Methods of Applying Urethane Lacquer Coating to Urethane Microcellular Foam

A variety of techniques can be used in coating molded urethane microcellular parts with the urethane lacquer. The parameters to be considered for application

Table 4. Tensile Strength and Elongation Properties of Urethane Lacquer After Exposure to Elevated Temperatures, 3 mil Clear Film.

Exposure Time, hrs.	Temperature, °F	Tensile Strength, psi Maximum	Tensile Strength, psi Yield	Elongation, %
0.5	176	5100	1820	365
18.0	176	4835	1685	345
0.25*	230	4670	1800	325
0.25*	266 **	4900	1770	345
0.25*	302 **	4785	1740	340
0.25*	338 **	4765	1720	345
0.25*	374 **	3774	1264	308

*Prebaked 0.5 hours at 176 °F.
**Film discoloration noted.

Table 5. Typical Film Properties for the Urethane Lacquer.

Specific Gravity	1.12
Hardness (Sward Rocker)	18
Tensile Strength, psi, maximum	5000
psi at yield point	1450
psi at 300% Elongation	3480
Elongation, %	350
Abrasion Resistance (Taber CS-10 wheel 500 gram load, 250 cycles, mg lost/1000 cycles)	5

of the urethane lacquer include:

a) mold size and configuration
b) material of construction
c) type and removal of release agent
d) in-mold or out-of-mold primer
e) surface texture required

Film-Foam Interaction

The microcellular foam composition, the urethane lacquer composition and the method of application play an important role in achieving a completely satisfactory part.

Good adhesion of the film coating to the microcellular foam is imperative. Proper release agents, surface treatments and foam formulation are also important aspects.

Surface defects in the molded part can occur from imperfections or foreign particles in the mold. Surface imperfections or bubbles may also occur in the foaming operation due to non-uniform mixing or improper formulation. Attempts to paint an imperfect microcellular foam surface will generally result in an imperfect finished surface.

The intrinsic color stability of the finished coated part is excellent. In some instances, improper formulation or non-uniform mixing of the foam can cause migration into the coating and ultimate discoloration. In addition, the composite urethane system of microcellular foam and coating must have tensile strength and elongation characteristics that are compatible to withstand severe impact.

URETHANE ENAMELS

Urethane coatings of the one-package, bake-enamel type depend upon the reactions of pendent groups on the urethane polymer and reactive sites available from another species of film-forming resin. The reaction between two different types of film-forming materials must be essentially nonexistent at normal storage temperatures for adequate package stability. On the other hand, film-forming reactions must proceed at a satisfactory rate at elevated temperatures ($>220°F$) to provide practical cure cycles.

A new color-stable urethane bake-enamel system under development by Allied Chemical has been found suitable for exterior microcellular foam bumpers. The new enamel system, based on a urethane polymer crosslinked with a melamine-formaldehyde resin, has the following characteristics:

1. One-package enamel
2. Excellent color stability
3. High gloss
4. High spray solids for application
5. Abrasion and impact-resistance

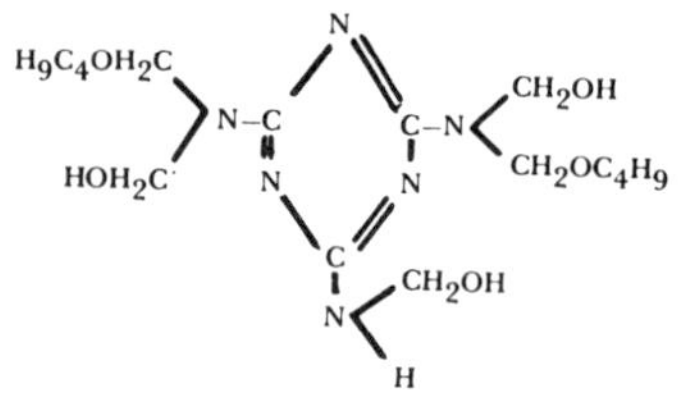

M. W. 320

Figure 5. Typical Structure of Melamine-Formaldehyde Resins.

6. Resistance to gasoline and oil

The urethane enamel requires a 220-250°F bake cycle for 30 minutes. A high gloss finish, without use of a clear coat, can be achieved with a variety of colors.

Cure Mechanism

The cure mechanism for film formation in the urethane bake enamel system is complex. The basic components are a hydroxyl terminated urethane polymer and a melamine formaldehyde resin.

The possibility of the melamine resin condensing with itself, as well as reacting with the pendent hydroxyl groups of the urethane polymer, must be considered. The competing reactions that are possible during the film-forming stage are schematically presented as follows:

These reactions indicate a wide variety of possible combinations in view of the large number of melamine formaldehyde resins and the wide latitude available for urethane polymer design.

Typical melamine resin compositions useful for urethane bake enamel systems are:

Uformite MM-83	(Rohm and Haas)
Uformite MM-47	(Rohm and Haas)
Cymel 300	(American Cyanamid Co.)
Cymel 301	(American Cyanamid Co.)

Figure 5 illustrates a typical structure for melamine-formaldehyde resin [8].

Typical solution characteristics for the urethane polymer are 1000 cps at 25°C, 50% N. V.

In some instances where temperatures of 250°F or less are required, catalysts should be added to enhance curing of the bake-enamel system. Typical catalysts (at concentration of 0.1 - 1.0%) are:

1. p-toluene solfonic acid
2. phenylacid phosphate (Mobil Chemical Co. PA-75)
3. Polycat 200 (Abbott Chemicals)

A wide variety of solvents and solvent blends can be used to reduce solids content suitable for spray application. Xylene, toluene, methyl ethyl ketone, methylisobutyl ketone, and butanol can be utilized.

Film Properties

Clear free films based on the urethane enamel were prepared in the same manner as described earlier for urethane lacquers. The urethane enamel system, based on experimental Urethane Polymer X-5128 P (70%), Uformite MM-47 (30%), and p-toluene sulfonic acid (1%), yielded a clear free film with a tensile strength of 3440 psi and an elongation of 32%.

Pigmentation of the urethane bake-enamel system is accomplished by conventional dispersion or grinding techniques. The pigment is generally introduced into the urethane polymer prior to the addition of the melamine resin. A typical enamel pigmented with duPont's R-900 (TiO_2) was sprayed at 29% N. V. on a prime-coated microcellular foam. Curing was effected at 250°F for 30 minutes. The resulting coated foam sample passed a 160 inch/pound impact test without cracking or wrinkling. The 60° gloss was 93. After 400 hours' exposure in a Weatherometer (WXR 150°F black plate temperature) the coated foam sample still passed the 160 inch/pound impact test and had a gloss of 68.

SUMMARY

In summary, two basic urethane one-package coating systems suitable for microcellular foam bumpers have been described. The mechanism for curing the urethane lacquers and the urethane enamel system illustrates a distinct contrast. The characteristic requirements for a coating system suitable for microcellular automobile bumpers was detailed.

REFERENCES

1. *Modern Plastics,* October 1967, page 92.
   Rubber Age, September 1967, page 83.
   W. W. Fitzgerald, P. A. Haines, E. P. Harris and R. F. Kienle, U.S. Patent, 3, 493, 257.
2. H. L. Ward, "The Automotive Market for Flexible Foams." Presented at The Society of the Plastics Industry, Cellular Plastics Division Technical Conference, February 24, 1970, Detroit, Michigan.
3. M. E. Bailey, M. Kaplan, B. Taub and G. S. Wooster, J. Cellular Plastics *5* No. 4, 237-46 (1969).
4. "Test Methods and Techniques for the Surface Coatings Industry" Shell Chemical Company, New York, N. Y. Second Edition, 1967.
5. S. Bonotto, and R. K. Walton, *Modern Plastics* Vol. 40, No. 9, 143 (1963).
6. G. C. Shaurr, T. K. Hay and M. Van Loo, *J. of Paint Tech., 38,* 501, October 1966.
7. R. R. Myers and J. S. Long, "Film-Forming Compositions" Volume 1, Part II, Marcel Dekker, Inc., New York 1967.
8. H. F. Payne, "Organic Coating Technology", Volume 1, John Wiley and Sons, Inc., New York 1954.

M. KAPLAN, *Specialty Chemicals Division, Allied Chemical Corporation*

CHAPTER 9

FLAME RESISTANT RIGID FOAMS

INTRODUCTION

Flame resistant is an uncertain designation. The term has been frequently applied to materials which are difficult to ignite, to materials which burn only when an outside flame is applied, and to materials which burn slowly. Most flame resistant materials do burn, however. Stone and glass, on the other hand, do not burn but cannot be considered entirely flame resistant when the thermal shock of flame temperatures causes cracking and other changes in physical form.

Flame resistance is a confusing subject which is affected by political and sociological as well as scientific considerations. This discussion is, therefore, a general analysis of material, and particularly rigid foam flammability and of the important factors which affect performance and testing.

MODELS OF COMBUSTION

The general process of combustion begins with an external heat source which raises the temperature of the material surface and produces a temperature gradient within the bulk of the material. When the temperature becomes high enough, the material decomposes, forming both gaseous and solid products. The flammable gaseous products combine rapidly with atmospheric oxygen, producing flame and heat to further raise the material temperature and continue the process. Solid products of decomposition form a char which is oxidized more slowly. The products of combustion diffuse out of the burning material while air diffuses in, and, if the amount of heat liberated is sufficient, the zone of combustion moves into the material, creating char, heat, smoke, and fumes.

In chemical terms, the combustion process occurs in the three stages shown in Figure 1. The flammable material is degraded by heating in stage (1). The flammable gases consist of both stable compounds and of free radicals, both of which are rapidly oxidized in the gas phase free radical reactions of stage (2). This stage produces flame and generates most of the heat required to continue combustion. The carbonaceous char which remains is oxidized more slowly in stage (3). The reaction occurs on the char surface with liberation of additional heat.

In the absence of oxygen, a similar process is known as ablation. The heat, in this case, must be entirely supplied by an external source, for example, air friction in the case of heat shields used for re-entry of space vehicles.

The heat is absorbed by the decomposition process and carried away by the gaseous products and radiation.

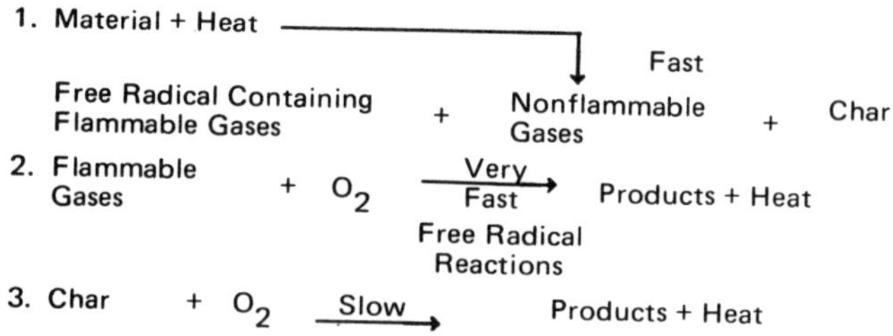

Figure 1. Generalize Mechanism of Combustion.

On the basis of this model, the flammability of any material is dependent on:

1. Heat transfer through the material.
2. Ease of decomposition to flammable gaseous products.
3. Availability of oxygen.
4. Heat of combustion (although heat evolved per unit volume may be more significant than the usual heat evolved per mole).

Examining the combustion of rigid foams in light of each of these factors, several differences from the case of solid materials result from the expanded structure. First, the heating effect of a given source will penetrate less rapidly into the material because of the lower thermal conductivity of most foams. Second, the gaseous thermal decomposition products are diluted by the gas contained in the foam cells. If this gas is air as it is in the case of open cellular and readily permeable foams, flammable gases burn more readily if their concentration becomes high enough. If, however, the cells are closed and contain a nonflammable gas such as $CFCl_3$ the flammable gases are diluted, air is excluded, and combustion is retarded. Third, the volume of combustible material per unit volume of foam is much lower in a foam than in a solid material and the available heat of combustion per unit volume is lower. Fourth, the open structure of a foam improves the air supply if the cells are interconnected or become so during combustion.

These considerations suggest that a foam should be less flammable than the comparable solid material if the charred foam maintains its physical structure. If, for example, a rigid urethane foam forms a surface char without cracking, the low thermal conductivity of the foam and presence of a nonflammable gas in the cells inhibits further combustion. On the other hand, if the foam melts or the char cracks and exposes unburned material, combustion continues.

This discussion has emphasized the importance of thermal decomposition to flammable products as a factor in determining flammability. The importance is apparent in data reported by Madorsky and Straus [1] listing pyrolysis products of phenolic resin at various temperatures and shown in Table 1. Increasing temperatures converted a larger fraction of the sample to volatile products, many of which were flammable. Similar compounds were identified [2] among the products of pyrolysis of several rigid urethane foams at 500°C., Table 2.

Table 1. Combined Mass-Spectrometer Analysis of Fractions Obtained in the Pyrolysis of Phenolic Resin.

	Amount, %, in volatile products[b]			
Component[a]	At 360°C.	At 500°C.	At 800°C.	At 1200°C.
Hydrogen	--	--	3.6	5.6
Carbon monoxide	--	3.5	16.2	24.6
Carbon dioxide	0.5	5.5	2.7	2.1
Methane	--	4.3	12.6	9.0
Acetylene	--	--	--	2.8
Ethylene	--	--	1.5	2.4
Acetone	6.7	17.6	1.0	--
Propylene	4.0	--	1.0	3.4
Propanols	10.9	11.1	--	--
Butanols	2.9	--	--	--
Cyclopentadienes	--	--	--	3.1
Benzene	--	2.5	0.6	2.8
Toluene	--	4.7	0.5	--
Dimethylbenzene	--	0.9	--	--
Others (not shown)	--	0	0.7	6.6
Vpyr	75.0	49.9	59.6	37.6
Total	100.0	100.0	100.0	100.0
Volatilized %	11	28	44	48

(a) Components in amounts of less than 0.5% are not shown.
(b) Percentages are based on weight of total volatilized part.

Table 2. Volatile Degradation Products Detected in the Pyrolysis of Rigid Urethane Foams at 500°C.

CO	Ethylene
CO_2	1-propene
H_2O	1-pentene
Benzene	Acetone
Toluene	Ethanol
Xylenes	2-propanol
$CFCl_3$ (Blowing Agent)	HCN
Chloroethane	HCl (From Flame Retardant)
$CHFCl_2$	
Chloroform	

The dependence of flammability on the formation of such products was shown by investigations of a series of isocyanate based polymers [3] of similar molecular structure. The flammability of the polyurethanes, polyureas, polyamide, and polyimide which were studied corresponded in every instance with the temperatures of initial TGA weight loss shown in Table 3.

FLAMMABILITY TESTING

The problem in flammability testing of any material is to determine what aspects of flammability are being tested and what the test results mean. No single test can measure flammability in a way which is meaningful for all applications and many tests correlate with others only with difficulty. Most organizations concerned with flame resistance have developed a combination of several tests for use in evaluating materials, but the selection is highly subjective. The number of tests is so great and the meaning of results so ambiguous that only a few standard tests are accepted for general specification purposes.

Table 3. Comparison of the Flammability and Thermal Decomposition Temperatures of MDI-Based Polymers.

Coreactant	Structure of repeating unit	Temp., °C., of TGA weight loss in air of		Flammability
		20%	40%	
Hydroquinone		280	300	Burned
p-Xylene-diol	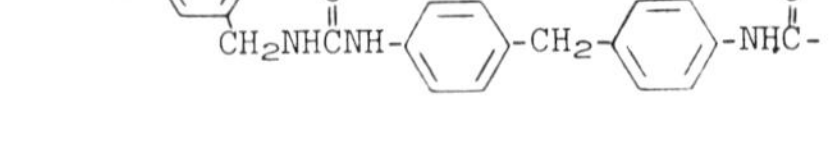	310	340	Burned
p-Phenylene diamine	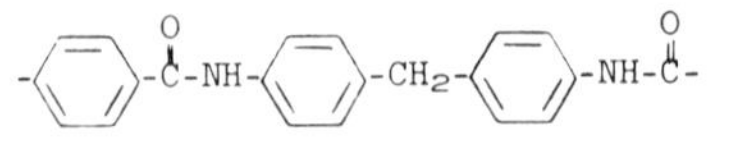	320	350	Burned
m-Xylene diamine	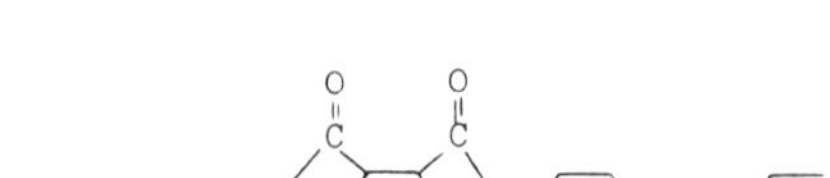	360	390	SE
Terephthalic acid	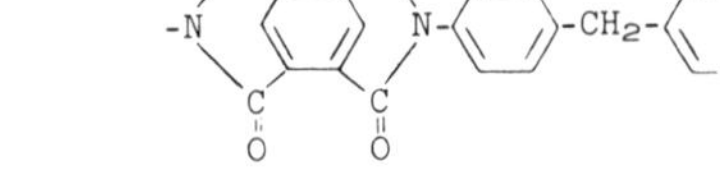	380	580	SE
Pyromellitic acid		560	605	SE

The tests in common use measure one or more of six properties.

1. Ease of ignition: Tests involve exposure of the sample to a controlled flame, a heated object, or a hot atmosphere. In the oxygen index test [4], the amount of oxygen in the atmosphere surrounding the sample is varied. Ratings are based on the start, continuation, and extent of combustion.
2. Flame propagation: These tests measure the rate at which combustion moves across the test sample.
3. Heat evolution: The amount of heat evolved by combustion of the test sample under controlled conditions is determined by calorimetric measurements or by temperature measurements at specified locations in the test equipment.
4. Endurance: The time for which a test sample will remain physically intact is determined, usually by exposing one side to flame or high temperature, then measuring the temperature on the other side or noting the time required for flame to penetrate the sample.
5. Changes of state and shape: Softening, melting, and distortion of materials exposed to fires are noted in all tests.
6. Products of combustion: The formation of the particulate matter which we call smoke and of invisible products such as CO, $HC1$, and SO_2 are measured by optical and chemical methods. These products both obscure visibility and produce toxic and corrosive effects.

The variations in test results are caused by basic differences in the physics of the methods.

1. Sample Size: The total amount of heat liberated increases with the sample size. Under most test conditions, the concentration of heat and, therefore, the temperature is greater for a larger sample.
2. Test geometry: Samples are tested in either vertical or horizontal positions and at angles between. Igniting flames are applied to either top or bottom surfaces.
3. Air flow: Some tests are conducted in a still atmosphere and depend on thermal convection for air circulation. Others specify a forced draft.
4. Heat flux: The radiant heat flux supplied to the sample surface increases the rate of combustion. Some tests, for example, the ASTM E-162 radiant panel, heat the entire sample surface by means of a source separate from the igniting flame. In others, a chimney geometry radiates heat from one burning surface to an opposite surface.
5. Flame form and temperature: Test flames vary in both shape and temperature. Sharply pointed high temperature flames are used in some penetration tests. Other tests are based on large, lower temperature flames such as are produced by freely burning gas or oil. Flame temperatures also vary with the fuel mixture used.

The complexities in combustion testing make it imperative that tests be chosen to indicate the flammability characteristics which are important in a given application. If a rigid foam is to function as a fire barrier, for example, a fire endurance test should be conducted. If the material is to be used as insulation or for general construction purposes, flame propagation is of great concern and determination of the Flame Spread Factor by the ASTM E-84 Tunnel Test is often required. Testing of materials for aircraft use commonly includes a determination of the concentrations of toxic fumes. Materials used in conjunction with electrical equipment must not produce corrosive fumes such as HC1. Large numbers of flammability specifications have been written by government agencies and industrial purchasers. They should be consulted by those interested in particular applications.

A more extensive discussion of flammability testing has been written by C. J. Hilado [5].

FLAME RESISTANT RIGID FOAMS

Many materials have been converted to rigid foams and of these, a large fraction either are flame resistant or can be made flame resistant by special additives. The foams are of several basic types.

1. Inorganic foams such as foamed glass or foamed aluminum which are nonflammable but which do melt at high temperatures. The manufacture of these products is difficult because of heat removal and only slabs of limited size can be produced.
2. Organic thermoplastic foams which, like poly(vinyl chloride), have flame retarding elements attached to the main polymer chain, or, like poly(styrene), can be made flame resistant by addition of chlorine, bromine, and phosphorus compounds. These foams have excellent properties for many applications. Low softening points have been raised by crosslinking in the case of foamed poly(vinyl chloride). Formation of HC1 during combustion of the chlorinated polymers can cause corrosion of electrical equipment and has limited some applications. Because production of thermoplastic foams requires special heating and molding equipment, they are not well suited for foam-in-place applications.
3. Thermoset foams include the common foamed phenolics, epoxies, and urethanes, all of which are prepared from low molecular weight resins and can be easily handled in foam-in-place operations. The phenolic foams, because of tight crosslinking and highly aromatic structure, are inherently flame resistant. High crosslink density also produces a friable material. The foamed urethanes and epoxies have less crosslinking and better mechanical properties, but are more flammable. Addition of phosphorus and halogen containing additives produces highly flame resistant products.
4. Inherently flame resistant organic polymer foams based on aromatic and

Table 4. Flame Resistant Rigid Foams.

Material		Flame Resistance	Mechanical Properties	Decomposition or Softening Temperature	Method of Application	Degradation and Combustion Products
Foam Glass	Inorganic	Excellent	Good	Very high	Slab	None
Concrete	Inorganic	Excellent	Good	High	Slab, in-place	None
Poly(vinylchloride)	Thermoplastic	Good	Good	Low	Slab	HC1 is corrosive
Crosslinked	Thermoset	Good	Good	Moderate	Slab	HC1 is corrosive
Poly(styrene)	Thermoplastic	Good with additives	Good	Low	Slab	Heavy smoke
Fluorocarbons		Good	Fair	High*	Slab	F compounds
Poly(organosiloxanes)	Silicone	Fair to good	Fair	High*	Slab	
Phenolic	Thermoset	Good	Friable	High*	Slab, in-place	Organic Fragments
Epoxy	Thermoset	Good with additives	Good	Moderate*	Slab, in-place	Organic Fragments
Poly(urethanes)	Thermoset	Good with additives	Good	Moderate*	Slab, in-place	N Compounds
Poly(isocyanurate)	Thermoset	Good	Friable	High*	Slab, in-place	N Compounds
Poly(imide)	Polycyclic	Good	Fair	High*	Difficult to process	
Aromatic Polyamide	Polyaromatic	Good	Fair	High*	Difficult to process	
Aluminum	Metallic	Good	Good	High	Slab size limited	
Cellulose Esters	Cellulosic Thermoplastic	Poor	Good	Low	Slab	

*Decomposes below softening temperature

other cyclic structures have been developed. Poly(imides) are used for high temperature electrical insulation and have been prepared in cellular form [6]. Aromatic poly(amides) are the basis for flame resistant fibers such as Nomex [7] and poly(amideimide) rigid foams have been developed [8]. Much attention has also been given to poly(isocyanurate) foams prepared by trimerizing polyisocyanates to an almost entirely cyclic structure [9, 10, 11, 12]. These foams are more difficult to prepare than the more common types noted earlier but the combination of high temperature and flame resistance is very attractive.

The important properties of a selection of rigid foams are summarized in Table 4.

RIGID URETHANE FOAMS

The rigid urethane foams are the most common of the flame resistant cellular building materials. From the standpoint of flammability, they can be classified in three groups.

1. General purpose: Products of this type contain no or few flame retarding additives and are used in applications where burning is unlikely.
2. Flame resistant: These foams contain sufficient amounts of phosphorus and halogen compounds to make them self-extinguishing, that is, a fire started by an outside source will cease when the source is removed.
3. Difficult to ignite: Foams of this type are modified by large amounts of flame retardant additives or by molecular structures which are inherently flame resistant.

The flame resistance of typical rigid urethane foams in each category is given in Table 5.

The urethane foams in Group (1) are usually based on branched polyether polyols and either toluene diisocyanate (TDI), and undistilled (crude) TDI, or a polymeric polyisocyanate made by phosgenation of an aniline-formaldehyde condensate. The polyether polyols are produced by addition of propylene oxide to glycerol, pentaerythritol, sorbitol, sucrose and other polyhydroxy and polyamino compounds. The aliphatic polyethers degrade readily when heated, giving low molecular weight fragments which contribute to flame ignition and propagation.

Foams based on TDI tend to be more flammable than those based on the polymeric isocyanates because the single ring aromatic radicals produced by degradation of TDI polyurethanes are more volatile than the branched ring systems of the polymeric isocyanates. With other variables constant, the polyfunctional ring systems of the polymeric isocyanates produce foams which are less flammable and give more char when burned [3, 14].

Group (2) foams are similar to those in Group (1) but contain additives which interfere with combustion. Most common are compounds of C1, Br, and P, all of which decompose during combustion to form Lewis acids which act as free radical scavengers in the reactions of stages (1) and (2) [14, 15]. The halogens are most effective when easily split from flame retardant molecules to form halohydrogen acids and halogen radicals. For this reason, aliphatic chlorine compounds are better flame retardants than the more stable chlorinated aromatics. Less stable aromatic bromine compounds are relatively effective, however.

The halogen containing flame retardants normally form volatile Lewis acids which act in the gas phase reactions of stage (2). They are, therefore, particularly effective in reducing flame spread. Phosphorus containing flame retardants decompose and form acidic oxyphosphorus compounds during combustion. Because these compounds are not volatile, phosphorus based flame retardants act mainly in the solid state, catalyzing char formation and reducing afterglow in chars when flames have been extinguished. The form in which phosphorus is added to a urethane foam has little effect on its effectiveness, in fact, red phosphorus has been shown to be a good flame retardant [16] at an optimum concentration of 1.5% by weight.

Most commercial flame retardants are liquids and can be handled easily. Because of the difference in the action of halogens and phosphorus, many additives contain both elements. Tris(chloroethyl) phosphate and tris(2,3-bromopropyl) phosphate are commonly used nonreactive flame retardants. A reactive additive, O,O-diethyl-N,N-bis-(2-hydroxyethyl) aminomethylphosphonate (Fyrol 6, Stauffer Chemical Company) is also widely used in urethane foams. Inorganic compounds such as ammonium phosphate are effective but are water soluble and less easily dispersed. A water insoluble ammonium polyphosphate, Phosgard P/30, (Monsanto Company) has also been used.

The addition of compounds which decompose on heating to give nonflammable gases has also been investigated. Ammonium carbonate, for example, decomposes to both ammonia and carbon dioxide which tend to dilute flammable gases and

Table 5. Rigid Urethane Foam Flammability.

	Rigid Urethane Foam		
Test Method	Group (1)	Group (2)	Group (3)
ASTM E-84			
Tunnel Test:			
Flame Spread Rating	1000	50-300	15-40
Fuel Contribution	500	100-500	15-50
Smoke Density Factor	500	100-500	15-50
ASTM D-1692 Rating	Burning	Self-Extinguishing	Self-Extinguishing
Burning Length, Inches	2	2	0
U.S. Bureau of Mines Torch Penetration Test, Burn Through Time, Minutes	1	2-15	10-90

snuff out existing flames. One of the most spectacular examples of this flame retarding mechanism has been developed by NASA scientists [14]. The addition of large amounts of poly(vinyl chloride) or poly(vinylidine chloride) and potassium fluoroborate, KBF_4, generated large amounts of HCl and BF_3 during foam combustion. When foam of this type was applied to the cockpit of a fighter plane, the nonflammable gas generation was sufficient to extinguish a jet fuel fire.

Flame resistant urethane foams have also been produced by incorporation of flame retarding atoms in the base polyols. A well known series of hydroxyl terminated polyesters (Hetrofoam Polyester, Hooker Chemical Company) based on chlorendic acid and similar polyesters of tetrachlorophthalic acid are of this type. The chlorine substituents in these resins are more stable than in chloroaliphatic flame retardants and act by inhibiting combustion of the structures to which they are bonded. Propylene oxide adducts of phosphoric acid are also common, but serve only to introduce phosphorus into the polymer where it acts in the normal manner.

The group (3) rigid foams are not all actually polyurethanes but are prepared from polyisocyanates. Most are attempts to prepare molecular structures which will not burn even if flame retardants are absent. A series of patents assigned to the Standard Oil Company of Indiana e.g., [8] outlines preparation of poly(amideimide)s by reaction of trimellitic anhydride and polymeric isocyanates. The molecules are largely aromatic, thermally stable, and flame resistant. Other organizations, among them Farbenfabriken Bayer AG [9], BASF [10], General Tire [11], ICI [12], and Mobay [13], have disclosed foams prepared by polymerizing polyisocyanates to the cyclic trimer or isocyanurate. The linear condensation product, carbodiimide, has also been disclosed.

Foams of these types have not been fully developed for commercial use but have much promise for the future.

CONCLUSION

Concern with personal protection and air pollution dictate that rigid foam development will move toward greater flame resistance and less generation of smoke and toxic, corrosive products during fire exposure. The rigid foams now available, if carefully selected and used, are satisfactory for most purposes, but directions of current research will bring important improvements. The growth of rigid foam applications seems assured.

REFERENCES

1. S. L. Madorsky and S. Straus, *J. Research,* National Bureau of Standards *55,* 223 (1955).
2. Mobay Chemical Company, Unpublished data.

3. J. K. Backus, D. L. Bernard, W. C. Darr, and J. H. Saunders. *J. Appl. Polymer Sci. 12,* 1053 (1968).
4. C. P. Fenimore and F. J. Martin, Combustion and Flame, *10,* (2), 135 (1966); *Modern Plastics, 44,* 141 (Nov. 1966).
5. C. J. Hilado, I and EC Prod. Res. and Devel., *6,* (3), 154 (1967); Flammability Handbook for Plastics, Technomic Publishing Company, Stamford, Connecticut, (1969).
6. J. K. Finke and G. F. Wilson, Second SPI International Cellular Plastics Conference, New York, (Nov. 1968).
7. Nomex, Product of E. I. duPont deNemours, Inc.
8. H. E. Frey, U. S. Patent 3,300,420 (1967).
9. E. Windemuth, G. Braun, and P. Hoppe (to Farbenfabriken Bayer AG), British Patent 908,337 (1962).
10. H. Nohe, R. Plate, and E. Wegner (to BASF), Belgian Patent 657,835 (1964).
11. L. Nicholas and G. T. Gmitter, *J. Cellular Plastics 1,* 85 (1965).
12. I.C.I., Belgian Patent 680,380 (1966).
13. D. L. Bernard, A. J. Doheny and J. K. Backus, Canadian 842,308 (May 19, 1970).
    D. L. Bernard, J. K. Backus and W. C. Darr, Canadian 845,228 (June 23, 1970).
    D. L. Bernard and A. J. Doheny, Belgian 723,151 (Dec. 31, 1968).
14. J. A. Parker, S. R. Riccitiello, W. J. Gilwee and R. Fish, *Sampe J., 5,*41 (April-May, 1969).
15. J. H. Saunders and J. K. Backus, *Rubber Chem. and Technol. 39,* (2), 461 (1966).
16. H. Piechota, *J. Cellular Plastics 1,* 186 (1965).

J. K. BACKUS, *Mobay Chemical Company, Pittsburgh, Pennsylvania*

CHAPTER 10

DUROMER-AN INTEGRAL SKIN RIGID URETHANE FOAM

INTRODUCTION

The technology of molding cellular urethane plastics has advanced very rapidly during the past two years. Newly developed chemical raw materials and molding techniques now make it possible to produce finished parts with solid outer skins and microcellular interiors in a single operation. This urethane Duromer is in the class of structural or integral skin foams. These materials have a sandwich-like structure possessing a greater than normal stiffness at a relatively low weight. Contrary to conventional sandwich materials, the surface skins of Duromer parts are not formed from different substances but are urethane same as the core. They are formed simultaneously with the cellular interior from the same raw materials. The dimensions and properties of the skins can be controlled, to a certain extent, by suitable variations of chemicals and processing parameters.

The density distribution of the material over the cross-section of the molded parts is approximately parabolic. The solid surface skin with its high mechanical properties is transformed into the microcellular core continuously and the density decreases progressively toward the center of the cross-section. This type structure is very difficult to obtain in conventional sandwich materials.

Duromer shows many superior physical properties when compared with conventional high density, rigid urethane foams. Because of its physical structure, high strength and good dimensional stability at normal temperatures, Duromer should be very well suited to the production of relatively thin-walled articles with large surface areas. A product of such description might well fit into the construction, automotive or furniture industries. The latter industry seems very receptive to this new technology at the present time.

RAW MATERIALS

The starting materials for the production of Duromer are polyisocyanates and polyfunctional alcohols which are liquid at room temperature. These materials as well as stabilizers, activators and blowing agents can usually be preblended and used in the form of a two-component system that can be best handled through high pressure mixing equipment. Tertiary amines are most often used as catalysts for these formulations and by controlling the rate of chemical reaction affect the cycle times of the molding operations. Halogenated hydrocarbons such as monofluorotrichloromethane are the preferred blowing agents in these foaming systems.

PROCESSING

The processing of the liquid starting raw materials into finished Duromer parts follows a very well defined pattern. The following sequence of events takes place in making a finished item.

1. Chemical reactants are prepared by preblending to a minimum of two components for ease of handling in foam machines. The usual procedure suggests that polyol, blowing agent and catalyst form one component and the polyisocyanate the other.
2. The components are accurately metered and thoroughly mixed using a proper foam machine. The materials are mixed in a fraction of a second.
3. Mixed materials are injected directly from the machine into a closed mold cavity only partially filling it. This operation takes from 2 to 15 seconds.
4. Chemicals react expanding the polymer mass to fill the mold. This takes 30 to 90 seconds.
5. Foam then cures and partially cools to form a rigid item. Depending on many variables, this phase of production takes from 3 to 20 minutes.
6. Part is demolded.
7. Part is held for continued cooling.
8. Part is prepared for finishing.
9. Part is finished in a conventional manner.

In order to be assured of a high quality product, the foaming process must be measured and controlled. The terms involved with these measurements are a special language that has developed along with the urethane foam industry. A list of these terms and their importance follows.

1. Cream Time is the interval between the start of mixing of reactants and the color change of the mixture that denotes a start of bubble formation (foaming). This value is important since it limits the time that material can be injected into a mold.
2. Gel Time is that time when the foaming mass changes character from liquid to solid and is determined by the fact that the reacting polymer can be drawn from the mass into a fiber. This value shows the maximum time allowed for a reacting foam to fill a mold.
3. Demolding Time is self-explanatory and indicates that time at which a part has gained enough strength to support itself outside a mold. The demolding time is dependent on the raw materials and catalysts used, however equally important are part geometry and temperature cycle of the mold.
4. Free Rise Density is measured on a foam that is poured in an unrestrained manner and gives a good indication of proper formulation and machine operation. This value should be determined before a molding run is started.
5. Index Number refers to a correct chemical formulation. An index of 100 means that there is enough isocyanate in a formulation to react with all the reactive chemical groups in the polyol component.

This value is usually slightly greater than 100 so that one is assured of complete reaction, however it can be varied to control certain physical properties of Duromer and should not be changed from recommended ranges. Wide deviation can result in a very poor product.

Depending on the sophistication of the production facilities, the blowing agent will or will not be blended in the resin at the time of delivery. Assuming not, it must be added in a regulated amount, depending on the formulation used, before the resin is fed to the foam machine. During this preblending operation, it might also be necessary to adjust the catalyst level to obtain proper operating conditions for production. Care should be taken to be sure that a homogeneous mixture is obtained. Special attention should be given to the blowing agent which could easily remain in the bottom of the tank because of its higher specific weight. This could lead to bad parts. During these blending operations, air should have been dissolved into the resin component. This dissolved air is helpful in nucleating the foam formulation and producing a good foam structure. Finally, care should be exercised during the mixing operations, to make sure that water vapor is not added to the resin blend. The presence of water can cause serious foaming problems and result in unacceptable Duromer parts. Under ideal conditions, using proper equipment, the blowing agent and catalyst can be added to the prefoam mixture by using auxiliary pumping systems on the foam machine.

FOAM MACHINE

The use of a foam machine is absolutely necessary in the production of Duromer parts because the raw material systems react extremely fast. Processing by hand is not recommended even for experimental purposes since wrong conclusions may be drawn from poor data. Generally, the raw materials for Duromer can be processed using most of the known two component metering-mixing machines as long as the following conditions are fulfilled.

1. The mixing chamber volume should be kept as small as possible while still providing for good mixing.
2. The mixing chamber should not be sensitive to pressure variations.
3. The mixing head should be as light as possible and be connected to the metering machine by means of long flexible hoses for ease of handling and connecting to molds.

Easy attachment of additional pumps for the separate metering of blowing agents and catalyst into the resin component should also be possible. Machine metering of additives into the resin stream gives the option of controlling the foam formation within certain limits by variation of the blowing agent content and the activity of the mixture. The delivery rates of the metering pumps for the different components should be continuously variable, so as to give a freedom of choice in chemical reactants. The total output of the machine should cover a wide range so that the delivery times can lie between 3 and 15 seconds with no change in metering accuracy or great limitation of part weight. The raw material must be pumped in an exact ratio from the beginning to the end of the filling process without fore- and after-run. The metering accuracy of the pumps should be insensitive to pressure variations that occur during the processing, and finally it should be possible to incorporate the machine, complete with programmer, for different part weights, into an automatic production line.

The machines should have capacities that will allow the production of parts weighing from only a few grams to several hundred pounds. The Mobay HK—series foam machines provide the above listed performance characteristics.

MOLDS

The quality of molds for the Duromer process is just as important as for conventional injection molding operations. The mold surface is reproduced in exact detail by the Duromer part so scratches and pinholes on the mold surface show up on each part made in that mold. The pressures developed during foaming cannot be predicted in every case, but usually fall between 25 and 45 psi. Therefore, all molds and clamps should have a designed strength of about 100 psi for safety and performance reasons. These pressures are much lower than injection molding pressures, thus allowing a lighter mold construction. Care should be taken however, that there is sufficient stiffness to prevent bulging of the surface of large molds.

The formation of a solid skin on Duromer is not only dependent on the chemical formulation but also requires a specific temperature of the mold during foaming. For medium and large run production one must be able to control the mold temperature and this can be satisfactorily accomplished only with metal molds. The best temperature control can be obtained with a cooling-heating liquid being circulated through the mold, however, other methods such as oven heating and water spray cooling are also feasible.

RELEASE AGENTS

During reaction, Duromer like all other urethanes, goes through a tacky stage when it is possible that the part will adhere strongly to the mold. In some cases it is possible to demold without a release agent, however, it is usually recommended to decrease the adhesion forces and prevent mold damage as well as speed up cycle time. Many different release agents have been evaluated. Most of these are waxy products or metal soaps which are applied to mold surfaces in solid, paste or liquid form. For large production runs, liquid release agents are preferred because they can be applied quickly and evenly with the help of a spray gun. The large number of release agents on the market does not allow complete evaluation, so the user should test several products to find the one that gives the best results in his particular application.

The proper choice of a release agent involves the post treatment of the part as well as the release efficiency of the product. The adhesion of a coating is often adversely affected by the release agent making a thorough cleaning mandatory before painting is possible.

MOLDING TECHNIQUES

It is not possible to specify a filling technique that will work for every mold geometry; however practical experience has yielded the following general considerations.

1. The filling time depends on the size of the foam machine, the mold volume, the geometry of the part, the cross-section of the filling hole, the different flow cross-sections and the activity of the reaction mixture. The filling time (2-15 seconds) should be shorter than the cream time, because filling on top of a foaming mixture can cause material defects and density variations. As indicated before, there is an HK type foam machine of a proper size to meet these requirements for most practical applications.
2. The position of the mold during the filling operation is determined by the geometry of the part and the flow path and venting features of the mold. The final position of the mold is found by trial and error and must be such that air pockets are not trapped in the part. Parts with large surface areas and 3/8 inch to 3/4 inch thickness can be molded best in a vertical or slightly tilted (approx. 20°) position. The showing or important surface of the part should be made facing downward. Highly contoured molds should be positioned so that an even flow of the reaction mixture is obtained and proper venting achieved usually at the highest point of a cavity.
3. The quality of a part also depends on proper venting. A general rule is that a mold should be liquid tight but not air tight. The molds halves should match reasonably well. The first moldings will show where air accumulates and indicate where small vent channels should be placed in the parting surfaces. There is no general rule for size and number of vents, so this must be found for each individual mold.
4. It is important that the sprue be designed so that a turbulence free flow pattern is obtained. For larger parts, a sprue position below the liquid level has been found preferable for obtaining an even flow. Bubbles present at the start of the filling operation are at the top of the reaction mixture and can be removed through vents at the high point of the mold. Of course, the below liquid surface filling technique will require a proper valving arrangement to be used with the mixing head of the foam machine. When filling above the final liquid level, only a simple closing plug is required for the mold. This

second procedure often results in trapped air bubbles.

Since the production of good parts depends so much on experimentation, one must make the final adjustments of conditions by trial and error.

PHYSICAL AND CHEMICAL PROPERTIES

As stated before, a Duromer is not homogeneous, but has a sandwich structure. Therefore, it is not possible to apply the testing methods used for thermoplastics and homogeneous foams without modification, because they do not take into account the structure of this new material. For this reason, we have modified existing test methods so as to better show the properties of Duromer. The mechanical properties of Duromer are normally measured on specimens taken from sheet material, molded in a one meter square, 10 mm thick mold. The values reported here were obtained from these sample slabs, 10 mm thick specimens possessing two skins. Preparation of the test pieces was done using a high speed circular saw for bending rods and impact bars and with a high speed milling machine for tensile strength specimens. When the properties are reported together with the overall density, the density is measured directly on the test specimen.

The following Tables of data relate specifically to the Duromer systems listed. It can be seen that the strength and stiffness values of the different products vary with changes in chemistry.

As may be noted, compression strength properties were not included in the above list. These values would be controlled by the strength of the lowest density section of the specimen core and would vary widely within a single sample depending on thickness and other variables.

Experimental Product E 341 when foamed with Mondur® MRS produces a very reactive system which allows rather short mold cycle times. This cross-sectional parts can be demolded in less than five minutes if properly supported during the cool out period. This type Duromer material also has the best thermal stability characteristics of those now available. This product is dimensionally stable to nearly 120°C. Somewhat lower modulus is the major shortcoming of the material. Applications requiring good heat stability but only moderate structural performance would be filled by this product.

A typical formulation would show 110 parts of polyol and blowing agent as one component and 125 parts of Mondur® MRS as the other. The mixture would start foaming in about 15 seconds and gel in about 30 seconds. This polymer cures so rapidly that it can be demolded in less than five minutes even though still quite warm.

As indicated before, compression strength values were not included with the above data because of the possible great variation found in a single sample.

Experimental Product E 342 and Mondur® MRS produces one of the strongest products found thus far. This material has a high modulus but lower recommended use temperature and somewhat longer demolding delay

Table 1. Physical Properties of Duromer Based on Experimental Product E 341 and Mondur MRS.

Properties		
Specific Gravity, gm./cm.^{3}		0.65
Density, lb./ft.^{3}		40
Tensile Strength, psi.		2500
Tensile Elongation, %		6
Tensile Modulus, psi.		110,000
Impact Strength, ft.lb./in.^{2} (no notch)		5-10
Flexural Strength, psi.		6100
Flexural Modulus, psi.		120,000
Maximum Use Temperature, °C.		115-125
Hardness, Shore D		75-80
Thermal Coefficient of (-40 to +80°C.)		
Linear Expansion, in./in./°F.		3-5 x 10^{-5}
Thermal Conductivity, "k"		0.4-0.8
Water Absorption, wt. %		<1
Water Vapor Permeability gm./m^2/day		<1
Heat Reaction cal./gm.		45-50
BTU/lb.		80-90
Specific Heat, cal./gm.		0.33
Exothermic Temperature, °C.		150
Dielectric Constant,at 70°C.,	50 Hz	1.9
	800 Hz	1.8
	1 M Hz	1.6
Dielectric Loss Factor, 70°F.,	50 Hz	0.024
	800 Hz	0.007
	1 M Hz	0.016
Surface Resistance, ohms.		1 x 10^{11}
Volume Resistivity, ohm-cm		2 x 10^{14}
Dielectric Strength, kv./cm.		50

Table 2. Physical Properties of Duromer Based on Experimental Product E 342 and Mondur MRS.

Properties		
Specific Gravity, gm./cm.^{3}		0.65
Density, lb./ft.^{3}		40
Tensile Strength, psi.		3000
Tensile Elongation, %		6
Tensile Modulus, psi.		160,000
Impact Strength, ft.lb./in.^{2} (no notch)		5-10
Flexural Strength, psi.		7500
Flexural Modulus, psi.		170,000
Maximum Use Temperature, °C.		90-100
Hardness, Shore D.		80-85
Thermal Coefficient of (−40 to +80°C.)		
Linear Expansion, in./in./°F.		3-5 x 10^{-5}
Thermal Conductivity, "k"		0.4-0.8
Water Absorption, wt. %		<1
Water Vapor Permeability, gm./m^2 /day		<1
Heat of Reaction cal./gm.		45-50
BTU/lb.		80-90
Specific Heat, cal./gm.		0.33
Exothermic Temperature, °C.		150
Dielectric Constant, at 70°C.,	50 Hz	1.9
	800 Hz	1.8
	1 M Hz	1.6
Dielectric Loss Factor, 70°F.,	50 Hz	0.024
	800 Hz	0.007
	1 M Hz	0.016
Surface Resistance, ohms.		1 x 10^{11}
Volume Resistivity, ohm-cm.		2 x 10^{14}
Dielectric Strength, kv./cm.		50

of 5-8 minutes. It would be used in applications which require more structural involvement.

A typical formulation for this material shows 110 parts of polyol and blowing agent and 140 parts of Mondur® MRS. The time for foaming initiation is 20 seconds and the rise is complete after 50 seconds. The cure of this Duromer is somewhat slower and demolding should be delayed to 8 or 10 minutes. This type of Duromer possesses an excellent high density skin and molds at low pressure.

A useful characteristic of Duromer which contributes to its excellent molding performance, is the very low shrinkage during cooling. The shrinkage is usually between 0.2% and 0.5% depending on the cell structure. As a consequence of the very low shrinkage, it is possible to have adjacent cross-sectional thickness that vary considerably without noticeable "sink marks" on the reverse side of a very intricately shaped article. It is the low viscosity and good flow characteristics of Duromer systems that make it possible to produce these very intricate articles that might not be possible using the other molding processes.

Duromer is also resistant to most chemicals and similar materials that might come in contact in use. These materials include water, salt water, aliphatic hydrocarbons and a large number of other solvents. Certain strong solvents can cause some swelling, however there is no known materials for dissolving Duromer.

The following Table shows the effects of long term exposure of Duromer to a series of different chemical agents.

Table 3. Effects of Chemical Exposure on Properties of Duromer.

Test Liquid	Temp., °C.	Wt. Change %	Thickness Change, %	Tensile Str. psi.	Elongation %
None	—	—	—	1850	6.0
Water	20	2.4	1.35	1470	6.4
	50	13.2	1.23	2000	5.9
Salt Water	20	13.9*	0.81*	1950*	4.4*
Furniture Wax	20	0.9	0.72	1480	7.3
Kerosene	20	0.5	0.30	1640	7.7
Fuel Oil	20	0.7	0.76	1590	8.2
Motor Oil	20	0.2	0.66	1930	7.4
	60	1.9	0.18	2290	5.3
Lighter Fluid	20	0.3	0.69	1470	6.8

All specimens had two sides where cut foam was exposed.
Specimens were exposed 6000 hours.
*Specimens were exposed 4500 hours.

It can be noted that there was very little property change at room temperature, however some swelling did take place in high temperature water.

Long term exposure (1200 hours) to a series of water solutions with pH values ranging from 1 to 13 had very little effect on the physical properties of Duromer. The test specimens all gained a little weight by absorbing water through the cut edges, however, the flexural strength in all cases dropped less than 15% and the flexural modulus appeared unchanged within experimental error.

The aging behavior of Duromer was evaluated in the Xenon chamber (UV), the Weather-O-Meter and by outdoor exposure. After 1000 hours Xenon illumination, a yellowing of the material was obtained, which did not, however, influence the mechanical properties nor decrease the scratch resistance of the surface. Outdoor exposure at the seashore caused a yellowing of the surface after one year exposure. After a few hundred hours in the Weather-O-Meter, the exposed surface became dull and simultaneously the scratch resistance decreased. The surface dullness could be removed by polishing. The other mechanical properties were not influenced even after 2000 hours exposure.

Lacquered samples were exposed for 7500 hours in the Weather-O-Meter. The adhesion of the coating remained perfect, even on previously damaged surface areas..

The expected behavior of Duromer under conditions of cyclic load application is of interest to most prospective users. Data of this type are very difficult to measure and require long testing periods. Many extensive tests plus the experience obtained from practical application have shown that the fatigue properties of Duromer were good when compared to other polymeric materials. When flex fatigue behavior is defined as that ratio between the tensile load that can be held indefinitively under cycling conditions and the normal tensile strength, these materials show a value of about 0.35.

MANUFACTURING

Not all applications involve the use of a single molding but require its incorporation into larger more complex structures. Many manufacturing procedures require that the product performs like other materials that might be used. The following information relates to manufacturing procedures.

1. The cleaning and finishing procedures are similar to that of wood and other plastics.
2. Repairing of broken or imperfect parts is not difficult. Cracks or damaged areas of the surface can be closed up with polyester or epoxy repair pastes. In general all fillers and pastes used in the furniture industry can be used. The repaired areas must be subsequently sanded and coated.
3. Bonding Duromer to itself or other materials of construction is quite simple. The best results were obtained with reaction type adhesives such as epoxies and urethanes. Medium strength bonds result when contact adhesives, wet adhesives (PVA) or solvent base adhesives were used. These general results were found when both

skin sections and cut sections were bonded. When clean surfaces, free from mold release agents, were bonded with reactive adhesives, the breaking points were found to be in the material as often as the adhesive interface. A properly prepared joint is as strong as the Duromer itself.

4. Nail and screw holding characteristics of Duromer are important in many furniture applications. Duromer can be used like wood in this respect. The same procedures can be used, but it should be noted that Duromer does not have a grain so there is much less chance of splitting taking place. The screw holding power of Duromer was 20% better than redwood but only about 80% as good as fir and 50% as good as oak. The performance has been seen to be about equal to plywood. This property of course varies with the density of the Duromer sample.

   A stable holding test wherein fabric was attached to Duromer and several different types of wood showed that Duromer was 25% to 100% better than white pine depending on the density of the foam and thickness of the skin.

This information is far from complete, because of the newness of the technology; however, work is continuing on a broad scale and additional data will be generated at a rapid rate.

REFERENCES

1. H. Piechota, "Urethane Duromer-A New Urethane Plastic," Proceedings of 2nd SPI International Cellular Plastics Conference, November, 1968.
2. H. Piechota, "A New Polyurethane Structural Foam," *Kunststoffe,* Vol. 60, January 1970.

WILLIAM C. DARR, *Mobay Chemical Company, Pittsburgh, Pa.*

CHAPTER 11

RECENT ADVANCES IN FLEXIBLE URETHANE FOAMS

INTRODUCTION

Flexible urethane foams were introduced into this country in 1954. These were polyester urethane slab foams, based on the pioneering work of Bayer in Germany. The load carrying characteristics and resilience of these foams were notably different from those of latex foam, already well established in this country. Polyester urethane foams have found wide application in such uses as fabric lamination, but have never successfully invaded the cushioning and safety padding markets, by far the largest volume applications for flexible foams.

With the commercial development of polyester flexible foams, in which General Tire was a pioneer, urethane flexible foams were able to enter bedding, cushioning and other high volume markets for foam. Today urethane flexible foams hold the dominant position among flexible foams, as is graphically demonstrated in Figure 1.

Within the last few years a number of new developments have begun to be factors in the growth of flexible foam usage. These include integral skin foams, microcellular foams, and cold molded foams. In addition, there have been a number of improvements in foam mixing and handling equipment, and many refinements in slab and conventionally molded flexible urethane foams. New polyols, surfactants and additives have all contributed to these changes. It

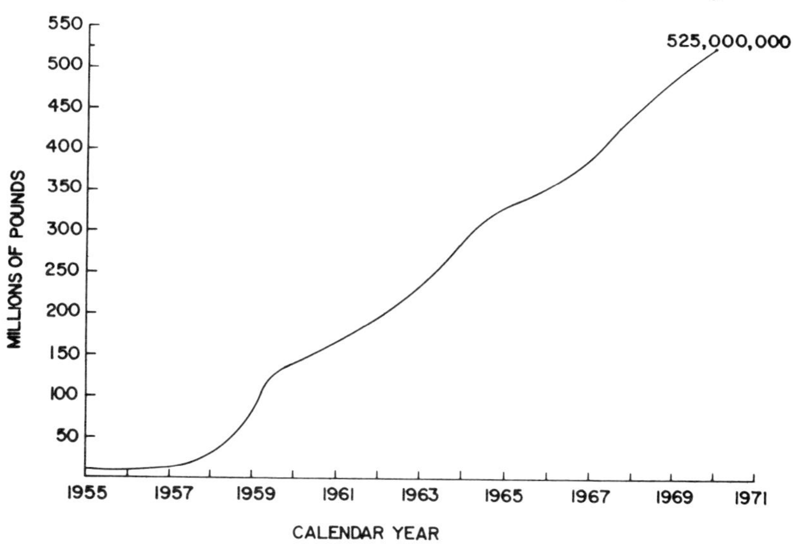

Figure 1. Production of Urethane Slab Foams.

Courtesy Mobay Chemical Company.

Figure 2. Pouring Slab Foam.
Courtesy Interscience Publishers.

Table 1. A Typical Polyether Urethane Slab Foam Formulation

Component	Parts by Weight
Polyol (Trifunctional)	100.0
Toluene Diisocyanate	46.0
Organo Tin Catalyst	0.4
Silicone Surfactant	1.0
Tertiary Amine Catalyst	0.2
Water	3.6
Monofluorotrichloromethane	0 - 15

Table 2. Properties of a Typical One-Shot Polyether Urethane Foam

Property	Value
Density (p.c.f.)	1.4
Tensile (psi)	14
Elongation (Per cent)	220
Tear (Lbs.,/In.)	2.2
Compression Set (Method B)	
At 50% Deflection	6
At 90% Deflection	5
Indent Load Deflection	
(Lbs. - 15 x 14 x 4 in. Sample)	
At 25% Deflection	30
At 65% Deflection	57
Ratio 65% Deflection/25% Deflection Load	1.9
Rebound Falling Ball (Per Cent)	38

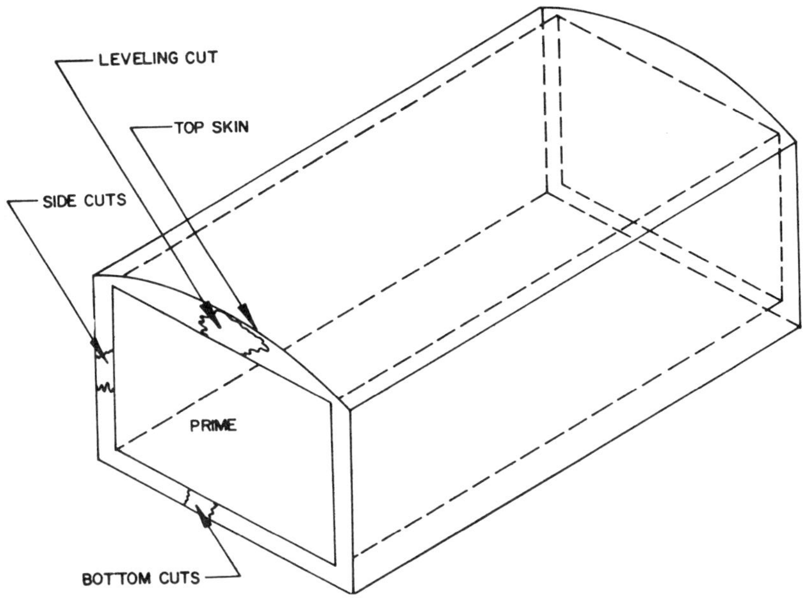

Figure 3. Configuration of a Slab Foam Bun.
Courtesy General Tire and Rubber Company.

is safe to say that more is happening in the area of flexible urethane foams than at any time since the late 1950's.

Historically speaking, flexible urethane foam was first available as slab foam. Slab foam is poured on a moving conveyor, usually lined with treated paper, and is "open blown," that is, unrestrained on top. Figure 2 conveys most essentials of the process. The mixer is on a traverse. As the conveyor moves the mix head moves back and forth, depositing the mixed foam components. A single pour may be hundreds of feet long, and have a cross section of seven feet by three feet or more, with a rate of pour in excess of five hundred pounds per minute.

The largest volume of flexible urethane foam produced is flexible polyether slab foam. A typical formulation is shown in Table 1, and the properties of such a foam in Table 2.

One result of this foam process method is a characteristic bun shape, shown in Figure 3. This results in a considerable quantity of foam poured which is not "prime," and must be further processed or sold for less than the prime material. Many efforts have been made to produce either a truly rectangular bun, or to reduce the crown effect as much as possible. In one of the earlier patents [1] paper in laid down on the foam material just before it starts to foam. In later patents, the side paper is made to move upward in relation to the foam froth, either by movement of the paper [2], or, very ingeniously, having a fixed ramp over which the bottom paper moves [3].

A rate difference thus occurs between side paper and foam rise.

Several problems are encountered in all these methods. Slab foam is poured commercially in various widths and heights, these being regulated by movable conveyor sides, material throughput and conveyor speed. Also, the expansion of the liquid material when pouring a 1.5 pound per cubic foot density foam, a typical density, is 45 fold, so that the foam in the froth stage before crosslinking has proceeded very far is extremely sensitive to pressure, vibration or shear such as induced by the rate differential of the side paper. It is obvious then that very sophisticated and flexible mechanical control is necessary. While various of these methods are alleged to be very effective, none has as yet been widely adopted.

The economics of slab polyether urethane foam production make it mandatory to have good control of the pouring operation. In order to hold a thirty inch high bun to a half-inch tolerance, for example, component temperature, metering and mixing and conveyor conditions must all be very carefully monitored. Properties, especially load carrying capacity, have very narrow tolerances as demanded by customers. In recent years much effort has been devoted to machine improvements.

Among such improvements in slab foam machines is a digital control of pump speeds [4], which in turn governs the amount of each component delivered to the mix head. Much study has been given to the factors governing fluid flow, especially for the more viscous polyol component.

The effects of mixer design on blending and nucleation of foams has received attention [5].

Equipment for handling the foam buns after pouring, storing them, and then slitting, cutting or otherwise converting the foam into the foam wanted by the customer, has become steadily more complex. Higher speeds and narrower tolerances have dictated this. In general, refinement rather than innovation has been the rule [6]. Figure 3 and Figure 4 are views of a recent slab urethane foam line, and illustrate the complexity of such an installation.

While almost all slab polyether foam is based on polyoxypropylene ether triols of 3,000-4,500 molecular weight as measured by hydroxyl number, and 80-20 isomer ratio toluene diisocyanate, primarily water blown, newer types of slab foam are making a bid for acceptance in the market place. These are known as high resilience or HR foams, based on high primary hydroxyl content polyether triols and polymeric isocyanates. One such formula is shown in Table 3 and its properties in Table 4.

This foam, while superficially similar to customary formulas, actually represents a very marked departure from the standard low density urethane flexible foams. One notable difference is in the lack of a metallo-organic catalyst. A still more surprising departure from usual practice is that this foam does not make use of a silicone surfactant. It depends on a very careful balance between rate of foaming and rate of crosslinking to maintain froth stability and thus produce a foam with a usable cell structure.

Figure 4. Pour Station-Flexible Slab Line.
Courtesy Interscience Publishers.

Table 3[a]. A High Resilience Urethane Flexible Foam

4,500 Mol. Wt. Polyoxypropylene Triol, 60 per cent Primary Capped	100
du Pont LD-813 Aromatic Amine	5
Dabco 33-LV Amine Catalyst	1
Water	2
du Pont Hylene TCPA Crude Toluene Diisocyanate	105 Index

(a) From Reference 7.

Table 4[a]. Properties of a High Resilience Urethane Flexible Foam

Density (p.c.f.)	3.0	
Tensile Strength (psi)	16	
Elongation, per cent	175	
Tear Strength, pli	1.8	
Compression Set, ASTM "B" 50% Deflection	<10	< 10
Indent Load Deflection 15 x 15 x 4 in. Sample lbs./50 sq. in.		
25% Deflection	24	
65% Deflection	63	
Sag Factor	2.62	
Static Fatigue - Method A	0.2	
Per Cent loss - 1 lb. ht.	0.2	
Per Cent loss - 25% ILD	20	
Ball Drop Resilience	69	
Flame Resistance (ASTM D-1692)	Self-extinguishing	

(a) From Reference 7.

Figure 5. Automated Control Station - Flexible Slab Line.
Courtesy Interscience Publishers.

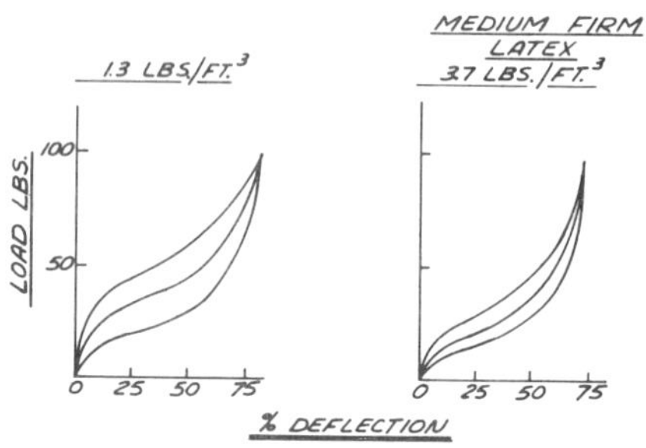

Figure 6. Stress-Strain Characteristics of Urethane and Latex Foams.
Courtesy E. I. du Pont de Nemours Company

Because of the markedly different polymer structure, foams of this type have quite different cushioning properties from the typical urethane foam. The cushioning properties are closer to those of latex foam, regarded by many as the most desirable.

To illustrate these differences in cushioning properties, I would like to introduce the concept of the load curve. Load curves for a low density polyether urethane foam and a medium firm pin core latex foam are plotted in Figure 6, together with the unloading curve. Foam hysteresis is measured by the difference in areas under the load and return curves. If one accepts the concept that desirable cushioning is a foam with low hysteresis, concurrent high resilience, and the characteristic load curve of the latex foam, then the plot in Figure 7, comparing conventional and aromatic diamine containing

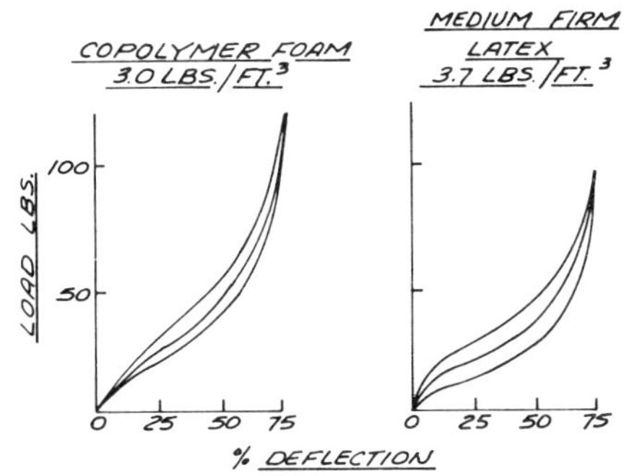

Figure 7. Stress-Strain Characteristics of Urethane and Latex Foams.
Courtesy E. I. du Pont de Nemours Company.

foams, leads one to the conclusion that the more novel formula is on a par with latex foam.

It has long been recognized that suitable formulations heavily loaded with inert fillers such as barytes, give foams which some feel are similar to latex foam in cushioning properties [8, 9]. A typical barytes loaded foam is listed in Table 5, and its properties in Table 6. A more recent filler for foams is a relatively large particle size silicous material, which its proponents claim gives higher resilience, lower hysteresis, and superior fatigue resistance.

NEWER USES OF POLYURETHANE FLEXIBLE SLAB FOAM

Slab foam is primarily used in cushioning, in furniture, bedding, and in recent years in steadily decreasing volume in automotive seating.

While the concept is not new, in the last year or two polyether urethane foam has made appreciable in-roads into the rug underlay market, displacing both sponge rubber and hair felt, which between them had monopolized this very large market. Urethane underlay is primarily either slit slab foam, suitably backed with film or non-woven fabric, or scrap slab foam rebonded in blocks and slit to appropriate thickness.

A new entrant into this market is slab foam, post-formed to higher density. While the process is proprietary and few details have been made public, this is an interesting approach to the underlay problem.

Notable among recent advances in polyester urethane foam production is the use of silicone surfactants designed for use in this type of foam. The wetting agents in use before the advent of the newer silicone surfactant have to be carefully conditioned before use and allegedly give less latitude in processing [10]. Equally interesting are the methods devised in order to pour cylindrical logs of foam. These efforts were motivated by the fact that most polyester foam is peeled to very thin gauges on veneer type peelers. A cylindrical log of foam gives maximum yield on equipment of this type. Three methods for pouring these logs have been publicized. One is the Scott process in which the foam is centrifugally cast in a rotating cylinder on a continuous basis. The Reeves and General Foam methods make use of specially designed conveyors. All of these processes are patented [11, 12, 13].

Reticulation of polyester urethane foam to give open pore material has opened new markets for the foam. The foam is converted by dissolving the membrane windows with caustic solution. Another method depends on permeating the foam with a flame explodable mixture of fuel and oxygen, exploding it to blow out the window membranes, and recovering the reticulated foam [14, 15].

One of the most interesting recent uses of polyester urethane reticulate foam is as a filler for gas tanks. In case of rupture or other damage, from impact or gunfire, the foam retains the gasoline or other fuel for a sufficient length of time to minimize fires. This use of urethane foam has been credited with saving many lives in Vietnam and in auto races.

Table 5. Higher Density Polyether Urethane Filled Foam Formulation[a]

3,000 Mol. Wt. Polyoxypropylene Triol	100.0
Barytes (Barium Sulfate)	100.0
Stannous Octoate	0.4
Silicone Surfactant	1.0
Triethylene Diamine	0.25
Fluorocarbon 11	0.15
Water	2.0
80/20 Isomer Ratio Tolylene Diisocyanate	31.0

(a) From Reference 8.

Table 6. Properties of a Filled Polyether Urethane Foam[a]

Physical Property	
Density, pcf	3.5
Tensile, psi	11
Elongation, Per Cent	182
Tear, ppi	1.4
Indent Load Deflection, (ILD) lbs./50 sq. in., 15 x 15 x 4 in. Sample	
25% Deflection	36
65% Deflection	80
SAG Factor	2.22
Static Fatigue 17 hours At 75% Deflection	
Loss in 25% ILD	21
Compression Set ASTM B	
50% Deflection	1.8
90% Deflection	4.7

(a) From Reference 9.

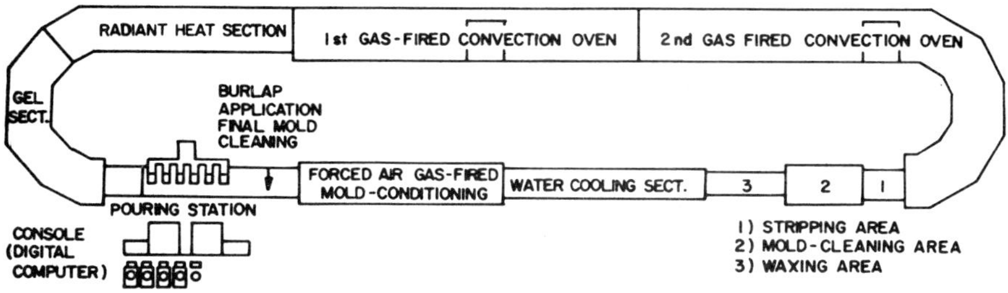

Figure 8. Layout of a Flexible Foam Molding Line.

Courtesy Modern Plastics.

FLEXIBLE FOAM MOLDING

Most molded flexible urethane foam is used in automotive seating. Topper pads, bucket seats and seat backs are typical products. In 1969, the great majority of urethane auto seats were molded, the remainder being die cut or otherwise fabricated from slit slab foam. By contrast the great majority of furniture uses of flexible urethane foam were based on slab foam.

The primary reason for the difference in use patterns is the cost of tooling, in the form of molds, so that a great many pads must be made in order to amortize this investment. Long runs of a given pad are the rule in automotive, the exception in furniture.

A modern urethane foam molding line must mold a diversity of complex shapes reproducibly with a low reject rate. The equipment to do this is necessarily elaborate and can represent a very large capital investment.

Using a schematic representation of a typical line, as shown in Figure 8, the process can be outlined by detailing briefly the passage of a mold through the various stations of the line [16]. This will emphasize some of the processing changes required to mold foam as opposed to slab foam manufacture.

The mold, coated with release agent and conditioned to a predetermined temperature, passes under the mix head to receive a precisely metered shot of the foam formulation programmed for this mold. The pour pattern is regulated by the speed of the mold as it passes beneath the head, the number of shots placed in the mold, and by traversing the head if necessary.

The mold now moves forward toward the curing ovens, and the mold lid closes. Normally, the mold passes into the curing ovens just after the end of the rise. Care must be taken not to jar or bump the mold at this stage, as the rising froth and initial set stages of the foam are sensitive to even small shock.

The mold then passes into a radiant heating oven of high capacity, where the mold is brought up to the exotherm temperature of the foam as rapidly as possible to replace the heat initially lost to the mold as a heat sink by the foam. This step is important in achieving good compression sets and good skin on the molded part.

In the next two oven sections the mold temperature is balanced out and then held at a temperature close to the exotherm temperature.

The mold now passes to the stripping station, where the mold opens, the pad is removed and placed on an auxiliary conveyor, where it is trimmed and inspected, while the mold is given an even cost of mold release.

The mold must now be cooled to pouring temperature, which in the machine represented by the diagram is done by cooled water jets. After this treatment the mold is conditioned to remove hot and cold spots. The conditioned mold then passes to the burlapping station. Burlap backing or fabric tape or other inserts, if called for, are affixed to the mold, usually to the lid, and after inspection and removal of any water or excess mold release, the mold is ready to repeat the cycle.

In order to run effectively the line, once started, must be held in a steady state with only small variations permitted in any step of the cycle. Shutdowns of more than a few seconds disturb the equilibrium of the cycle and may cause many defective pads before the balance is restored. To maintain this steady state both the chemical factors and the mechanical factors involved must be carefully controlled. As in slab foam processing, control of stream temperatures, metering, mixing, and pour pattern must be reproducible.

A mold line running only one or two types of pads, of simple shape and with a moderate depth of mold cavity represents the simplest situation which might be encountered. Here a relatively simple mechanical set-up could perform quite adequately. At the other end of the scale, a mold line running a variety of pad sizes and shapes, with some of complex configuration requires a very high degree of mechanical sophistication in order to perform adequately with a minimum of reject or repair pads. A graphic idea of the complexity of such a line is conveyed by two views of the console of a recent installation, as shown in Figures 9 and 10. In both instances, however, the same factors will be operating, so that the one situation is merely an elaboration on the same theme. Among the mechanical factors which must be considered, mold design, pour patterns, cure and mold release problems are among the most demanding.

Both radiant and high velocity gas fired ovens are used in the critical first stage heating. Both types have their proponents. After the first stage heating is complete, the ensuing ovens provide a "heat soak." Requirements of these ovens are, therefore, less demanding.

Because of the high amount of floor space, conveyor and energy requirements of these oven systems, with attendant expense, other modes of heating have been investigated. Of these microwave curing has received most attention.

In microwave curing the foam, in a mold transparent to this energy source, is raised to temperature by molecular excitation. The mold remains relatively cool, and the foam is very rapidly cured. Oven cycles as low as one minute have been reported [17].

The difficulties associated with microwave curing include problems in mold materials and construction and the problems associated with even

Figure 9. Control Panel Continuous Molding Line.
Courtesy Interscience Publishers.

Figure 10. Programmer - Continuous Molding Line.
Courtesy Interscience Publishers.

heating in the microwave unit without hot spotting. The possible savings in floor space and simplicity of this development are well displayed in Figure 11.

COLD MOLDING

The very elaborate installations needed for conventional urethane foam molding have inspired many efforts to devise a molding system which would not require the heating cycles and, therefore, the long conveyor lines and expensive ovens which we have shown are needed for TDI-water blown foams. These systems have been dubbed "cold molding" systems to differentiate them from the older systems. Where standard molding systems differ only in minor details from the formulation shown in Table 7, a number of significantly different systems have been proposed for cold molding use. Among these are the systems earlier discussed under slab foams [7], a system based on a proprietary quasi prepol [19], and a number of systems based on polymeric MDI [20, 21] of which the formulation shown in Table 8 is an example.

The common denominator in all these systems is a highly reactive polyether triol, of 4,500-6,500 molecular weight and over sixty per cent primary hydroxyls. In many, triethylene diamine (Houdry Dabco LV-33) is the catalyst, because of the unique reactivity of this tertiary amine. The physical properties of many of these systems differ markedly from the properties normally deemed acceptable for automotive molded pads, as shown in Table 9. On the other hand, they have cushioning properties very similar to latex foam, as shown in Figure 7.

Since these systems do not depend primarily on the use of metallo-organic catalysts or silicone surfactants to give acceptable foaming, but instead are controlled by viscosity, crosslinking rate and other factors less well understood, many have found that new concepts in equipment and mixing conditions are necessary. For example, a high rate of air injection has been found beneficial in producing moldings with good skin and uniform cell structure [19]. Slight over-packing of the mold has been recommended [21].

One approach uses a Hennecke HK machine as shown in Figure 12. This machine depends on impingement of two high velocity streams and does not use a mechanical agitator [22].

The availability of systems which do not require external heat, and can use inexpensive molds has inspired new approaches to seating, both automotive and informative.

As shown in Figure 13, large chairs and sofas, primarily foam, with or without an insert, are already commercial realities in Europe. The automotive industry, long wedded to "cut and sew" techniques for their seating, based on springs, with foam as padding and cut and sewn upholstery, is now seriously considering seating in which deep sections of foam in a plastic or metal shell, with, optionally, the foam already molded in vinyl or other upholstery material. Cold molding lends itself well to such a concept, since

Table 7. A Typical Flexible Polyether Urethane Molding Formulation[a]

Material	PBW
3,000 Mol. Wt. Polyoxypropylene Triol, Partially Ethylene Oxide Capped	100.00
Water	3.50
Fluorocarbon !!	10.00
DABCO 33 LV	0.45
N-Ethylmorpholine	0.60
Stannous Octoate	0.15
80/20 Isomer Ratio Tolylene Diisocyanate	44.90

(a) From Reference 18.

Table 8. A Cold Molding Formulation[a]

Materials	PBW
6,500 Mol. Wt. Polyoxypropylene Triol, Primary Capped	100
Water	4
N,N,N',N', Tetramethylbutanediamine	1
Polymeric MDI	50

(a) From Reference 20.

Table 9. Comparison of Properties of Conventional and Cold Molded Foams

Property	Conventional[a] Molded	Cold Molded
Core Density pcf	1.85	3.2
Tensile Strength psi	21	13.5
Elongation, Per Cent	290	124
Tear, ppi	2.5	1.3
Indent Load Deflection, Lbs. 50 sq. in. foot		
25% Deflection	16	23.1
65% Deflection	38	69.6
Load Ratio (Sag Ractor)	2.37	3.0

(a) Reference 20

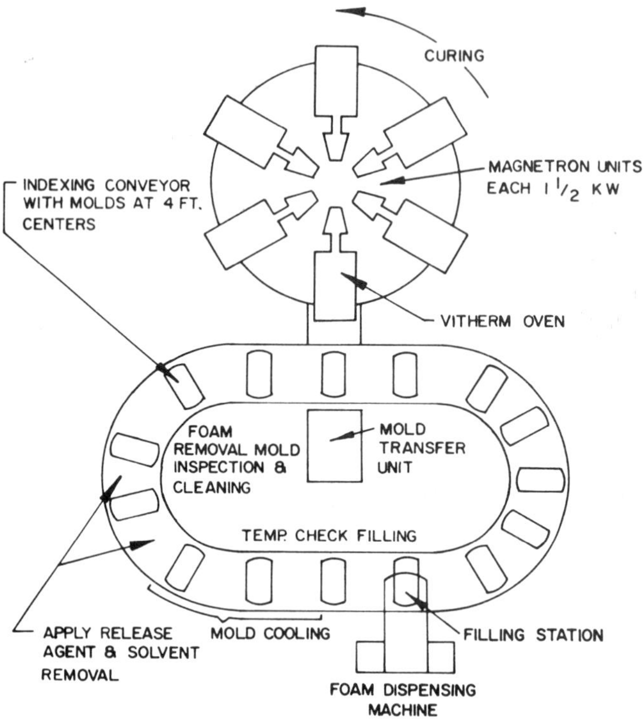

Figure 11. Layout of a Microwave Unit.

Courtesy Rubber Age.

Figure 12. Hennecke H-K Foam Machine.

Courtesy Mobay Chemical Company.

Figure 13. All Foam Furniture.
Courtesy Mobay Chemical Company.

the foam can be poured in a thermoformed vinyl or other skin. Also, the formulations previously discussed have excellent fatigue and height loss characteristics.

Such concepts, and others to be discussed, may involve significant re-orientation in foam use, with consequent changes in manufacture and marketing.

SEMI-RIGID FOAMS

With the government emphasis on auto safety, the use of semi-rigid urethane foams expanded rapidly. Mandatory crash pads, non-hostile arm rests, A-pillars, normally use semi-rigid foam in a vacuum formed cover of vinyl or other plastic.

The molding of semi-rigid parts is subject to many of the problems observed in flexible foam molding. While curing is not as difficult, the crosslink density is much higher in semi-rigids as is the density. The necessary good mold flow, coupled with rapid cure and open-cell structure to avert shrinkage are difficult to achieve [23, 24].

In order to obtain these balances, two approaches have been used. The first is the use of quasi-prepolymer systems based on polyoxypropylene triols of lower molecular weight with tolylene diisocyanate, water blown, with the crosslink density optionally increased by the use of high functionality, low molecular weight polyols. Another system now in use is based on high molecular weight triols, polymeric MDI, water or fluorocarbon blown. Again, lower molecular weight polyols of high functionality can be added to modify properties [25].

Table 10. A Semi-Rigid Foam Formulation[a]

	PBW
Quasi prepolymer*	226.0
1,000 Mol. Wt. Polyoxypropylene triol	85.0
700 Hydroxyl number amine based polyol	15.0
Water	4.0
Triethylene diamine	0.5
Calcium carbonate	30.0

*Quasi-prepolymer prepared from 1,000 mol. wt. polyoxypropylene triol and TDI, 15.1 per cent free NCO

(a) From Reference 23.

Table 11. Properties of a Semi-Rigid Urethane Foam[a]

Density, pcf	7.0
Ball rebound, %	51.0
25% CLD, psi	1.5
50% CLD, psi	-
50% Compression Set, %	5.8
Loss in 50% CLD, humid aging 5 hrs./121°C., %	68.0

(a) From Reference 23

Table 12. Integral Skin Foam Formulations[a,b,c]

Material	PBW
I	
4,500 Mol. Wt. Polyether Triol	100
Dimethyl Silicone Oil, 50 cs. Viscosity	1
Triethylene Diamine, 33% Solution	1.5
LD 813 Aromatic Diamine	20
Fluorocarbon 11	12
65-35 Isomer Ratio Tolylene Diisocyanate	20.5
II	
Niax 31-45 Polymer Polyol	85
Triethylene Diamine, 33% Solution	0.6
Dibutyltindilaurate	0.05
Fluorocarbon II	18
Silicone Surfactant	0.075
Isonol 93	15
Isonate 143L MDI	35
III	
Primary Capped Polyoxypropylenediol	79
1,4-Butanediol	11
Polyoxypropylene Tetrol. Mol. Wt. 450	10
Monofluorotrichlormethane	5
Methylene Chloride	2.5
Triethylene Diamine 33% Solution	1.5
Mondur MT 40	44

(a) From Reference 27
(b) From Reference 28
(c) From Reference 29

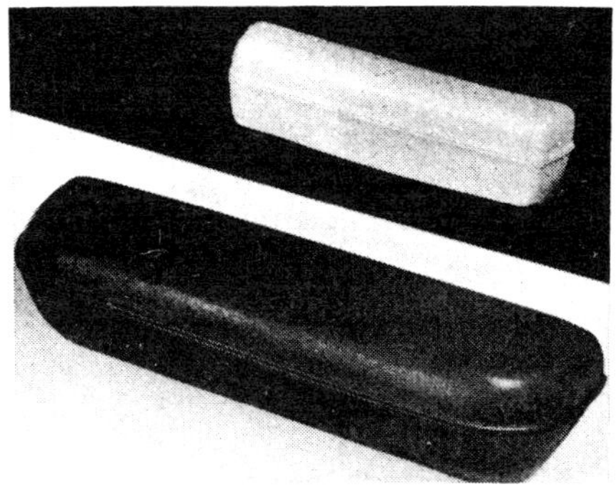

Figure 14. An Integrally Skinned Foam Part.
Courtesy Union Carbide Company.

A formulation of the first type is shown in Table 10 with its properties given in Table 11.

One outstanding advantage of formulations of this type, optionally extended with aromatic damines, is the very short molding cycles, e.g., five minutes, possible [25].

Another system of recent interest is one with a relatively flat response to temperature over a range from -40°C to 93°C [26]. Because of high crosslink density, many semi-rigid systems stiffen markedly at 0°C or less.

INTEGRAL SKIN MOLDING

The concept of an arm rest, as an instance, in which the urethane foam forms its own tough, decorative skin has come into prominence in the last year or two.

Many formulations have been suggested to accomplish an integrally skinned foam part. Table 12 gives an idea of the number of approaches being evaluated [27, 28, 29].

All of these formulations have one thing in common. All are solvent blown, in most instances, with fluorocarbon. This mode of foaming lends itself admirably to integral skin formation, since the foam expansion step is to a considerably degree dependent of polymer formation. As a result, a good deal of gas can be generated before the polymer growth has proceeded to a level where the evolving gas can be effectively trapped. The variations are in the polymer structure. Lacking the implicit urea groups which are invaluable in water blown foams to attain tensile and other properties, these must be achieved in other ways. In one, a hindered diamine (LD-813) is used, in another, short chain diols and a low molecular weight crosslinker, and one uses MDI to give a more symmetrical polymer structure.

All of these must be painted with a light stable coating, which may be based on light stable urethanes, or may be acrylic or other polymer with excellent stability to sunlight and other weathering conditions [30]. A part made by the use of this technology is shown in Figure 14.

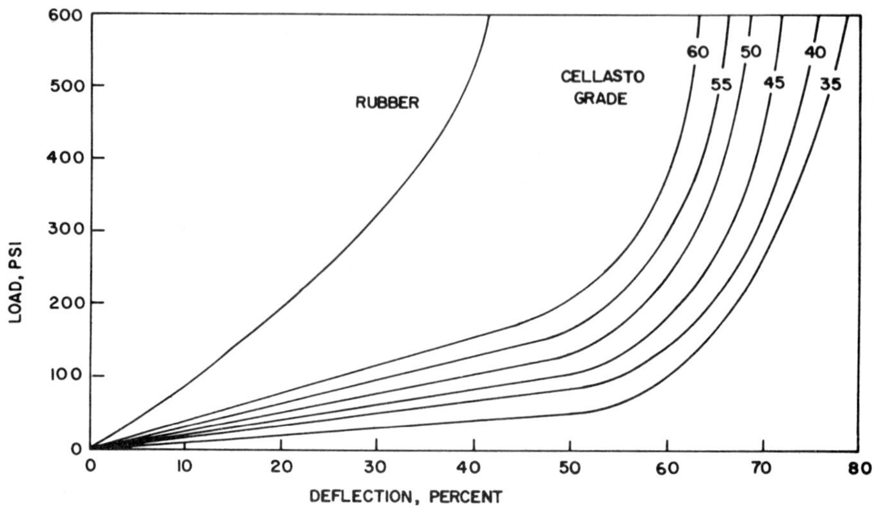

Figure 15. Load Curves of a Microcellular Foam.

Courtesy Mobay Chemical Company.

COMPOSITE SEATING

Almost from the earliest days of flexible urethane foam in this country, people have been intrigued by the possibility of producing in one or two steps a complete seat or upholstered cushioned furniture piece. Among early efforts was the Karpen chair, a joint effort of the Karpen Company and The General Tire and Rubber Company, Marion Division. The Bostrum Division of UOP has specialized in truck and other seating based on the idea of a composite. The continuing development of machinery and chemistry, several examples of which have already been given, have given new impetus to composite foam constructions.

Among recent examples of this trend are the Bostrum "Thinline" seating, Eaton Yale and Turner's spring and foam composite, and other examples already cited [31].

One of the major motivations for this trend is the continuing increase in the cost of labor to produce traditional seating. Additionally, these composites have unique characteristics of durability and appearance.

MICROCELLULAR FOAMS

By agreement, if not by precise scientific definition, microcellular urethane foams are urethanes, which have been foamed to densities of 15 to 60 pounds per cubic foot, and which are low enough in crosslink density to have good elastomer properties. Depending on the end use, they may be either open-cell or closed-cell foams [32].

The property of these foams of most interest at this time is the combination

Table 13. A Microcellular Foam Formulation[a]

Prepolymer of 80-20 TDI and Poly Tetramethylene Ether Glycol NCO, 6.3%	100.0
Methylene Chloride	8.0
Silicone Copolymer Surfactant	2.0
Cumene Diamine	5.5
m-Phenylene Diamine	3.9
Black Pigment	2.0

(a) From Reference 33 (Courtesy of Rubber Age)

Table 14. Properties of a Foamed Urethane Elastomer[a]

Density lbs./cu. ft.	25	37
Tensile Strength psi	500	900
Elongation Per Cent	400+	400+
Tear Strength psi	55	100
Compression Set at 70°C. and 50% Deflection	6.5	7.5

(a) From Reference 35 (Courtesy of Rubber World)

of high strength coupled with high shock absorbency possible with these foams. The load deflection curves shown in Figure 15 illustrate this.

The practical significance of these curves is that the foams, up to deflections of 30-40 per cent, deflect very readily, but beyond this point, the foams stiffen very rapidly. On impact, the material deforms readily at first without lateral distortion, and then stiffens very rapidly, so it is capable of withstanding impact forces as high as 5,000 psi while transmitting the impact to the associated structure at a lower rate and markedly cushioned.

Another unique property of the microcellular foam as compared with the solid elastomer urethane is the lower hysteresis. Open cell foams especially show low heat build-up under dynamic flexing at rates as high as 240 cycles per minute, while the solid elastomer often heats up to so high a level that it shows high property loss. Compressibility without lateral distortion up to quite high deflection is another property of these foams. Utilization of these properties has led to many applications, particularly in Europe, and many more are in process of development [32].

A number of systems have been developed for the formulation and processing of these foams. The formulation listed in Table 13 is one approach [33]. Typical properties are listed in Table 14.

Other systems are based on polyester urethane casting fluids, extended with 1,4-butanediol and blown with small amounts of water or with solvents such as monofluorotrichloromethane. These systems can also be extended with diamines or with polyols other than butanediol. One-shot systems based on either polyester or polyether polyols, with TDI or MDI as the diisocyanate and aromatic diamines or low molecular weight polyols as chain extenders are also available. Again, these can be water or solvent blown. A recent approach is the use of thermoplastic urethane elastomers compounded with suitable blowing agents and injection molded.

FLAMMABILITY OF FLEXIBLE URETHANE FOAMS

The increasing concern over flammability of materials, in which plastics and elastomers are often the whipping boy, has produced several recent developments in flexible urethane foams designed to reduce their tendency to support combustion.

One such development is the crude-TDI-hindered aromatic amine based foam cited earlier [7]. Another is the development of an isocyanate material which can be formulated into foams which pass currently recognized flammability test procedures [36].

Another approach is a new phosphorous-chlorine polymeric additive now available in pilot plant amounts, which is said to be non-fugitive and effective without undue loss of other properties required of urethane foams [37].

It must be remembered that flexible urethane foams, because of their low density and open cell structure, present one of the most exacting problems in the whole area of flammability. The validity of present test procedures is also undergoing careful scrutiny by workers in the field [38]. The continuing tensions due to these factors will, with time, produce a satisfactory and realistic appraisal of flexible foam flammability, but at present this has not been resolved to everybody's satisfaction [39]. For these reasons, the author anticipates a very lively period of research and development in this critical area.

CONCLUSIONS

From the many recent developments cited, some refinements in present techniques, some extensions of present technology, and some which can best be described as novel, it is obvious that flexible urethane foams, having solidly established themselves as useful and valuable materials, are in the throes of further expansion. It seems safe to predict that the phenomenal growth rate of flexible urethane foams will continue unabated. The exact direction of this expansion, with so many exciting possibilities is much less clear. It is clear that these foams can have a future at least as exciting as their past.

REFERENCES

1. Richard D. Joseph and Robert W. Mitten, U.S. Patent 3,354,503 to The General Tire and Rubber Company, 1967.
2. Derek Boon, U. S. Patent 3,325,828 to Reeves Bros., 1967.
3. Fred Duff, Marcus French and Warren Pollock, U. S. Patent 3,281,894 to General Foam Corp., 1966.
4. *Chemical Processing, 30,* No. 12, 137 (1967).

5. J. M. Bvist, *J. Cell. Plast. 1* 21 (1965).
6. Roger Knox, in *Polyurethane Technology* (Paul Bruins, ed.) Interscience Publishers, New York, New York 1969.
7. du Pont Technical Bulletin, *A New High Resiliency Flexible Urethane Foam,*1965.
8. Olin Mathieson Technical Bulletin, *Effect of Fillers in Urethane Foams,* 1965.
9. V. L. Currier, from *Proceedings of Polymer Conference Series on Cellular Plastics,* Wayne State University, Detroit, 1966.
10. Union Carbide Technical Bulletin, *A New Surfactant for Polyester Urethane Foams,* 1968.
11. Robert A. Volz, U. S. Patent 3,171,820 to Scott Paper Co., 1965.
12. F. Lorenz and G. Schlagenhof, U. S. Patent 3,297,803 to Scott Paper Co., 1965.
13. Fred W. Meisel and Edgar Allen Blair, U. S. Patent, 3,297,803 to Scott Paper Co., 1967.
14. William R. Powers, U. S. Patent 3,297,802 to Scott Paper Co., 1967.
15. Henry C. Green, U. S. Patent 3,175,030 to Chemtronics, Inc. 1965.
16. *Modern Plastics, 44,* 12 (1967).
17. J. M. Buist, *Rubber Journal, 147,* 26 (1966).
18. Union Carbide Technical Service Report, *Niax Polyols for One-Shot Closed Molded Foams,* September, 1964.
19. Union Carbide Technical Bulletin, *Cold Molded Urethane Flexible Foams,* 1968.
20. Dow Technical Bulletin, *Cold Molded Flexible Foams,* 1968.
21. Jefferson Technical Bulletin, *Cold Molded Flexible Foams Based on Polymeric MDI,* 1968.
22. Mobay Technical Bulletin, *Hennecke HK Equipment,* 1970.
23. Jefferson Technical Bulletin, *Urethane Chemicals for Semi-Flexible Foams,* 1967.
24. Hans Wirtz, from *Proceedings Second International SPI Cellular Plastics Conference,* New York, 1968.
25. R. B. Grant, from *Proceedings SPI Automotive Conference, February, 1970* Contribution No. 266.
26. Union Carbide Technical Bulletin, *Urethane Semi-Flexible Foams with Reduced Sensitivity to Temperature,* 1968.
27. du Pont Technical Bulletin, *New Formulations for Semi-Rigid Integral Skin Foams,* 1968.
28. Upjohn Technical Bulletin, *An Integral Skin Urethane Foam System,* 1968.
29. Mobay Technical Bulletin, *Integral Skin Foam Formulations Based on Mondur MR,* 1968.
30. Temple G. Patton, *Polyurethane Technology* (Paul Bruins, ed.) Interscience Publishers, New York, New York 1969.
31. Eaton, Yale and Towne, U. S. Patent Application.
32. R. C. Burck, from *Proceedings of Second SPI International Cellular Plastics Conference,* 1968.
33. B. Dianca and R. E. Knox, *Rubber Age, 98,* 76 (1966).
34. R. E. Knox, *Polyurethane Technology* (Paul Bruins, ed.) Interscience Publishers, New York, New York 1969.
35. R. P. Kane, *Rubber World, 147,* 35 (1963).
36. Union Carbide Technical Bulletin, *Niax Isocyanate FR for Flame Retardant Flexible Urethane Foams,* 1969.
37. Monsanto Bulletin, *A New Phosgard for Flexible Urethane Foams,* 1970.
38. C. J. Hilado, *Polyurethane Technology,* (Paul Bruins, ed.) Interscience Publishers, New York, New York (1969).
39. Irving N. Einhorn, *Proceedings of University of Detroit Flammability Symposium,* 1969.

E. M. MAXEY, *Research Associate, General Tire and Rubber Company, Akron, Ohio*

CHAPTER 12

RIGID URETHANE FOAM SANDWICH STRUCTURES

STRESS ANALYSIS

The basic stress analysis of foam-filled structures is characterized by the following:

1. Deflection is increased by shear deformation of the core.
2. Creep.
3. Buckling is affected by the rigidity of the core.
4. Thermal warp.

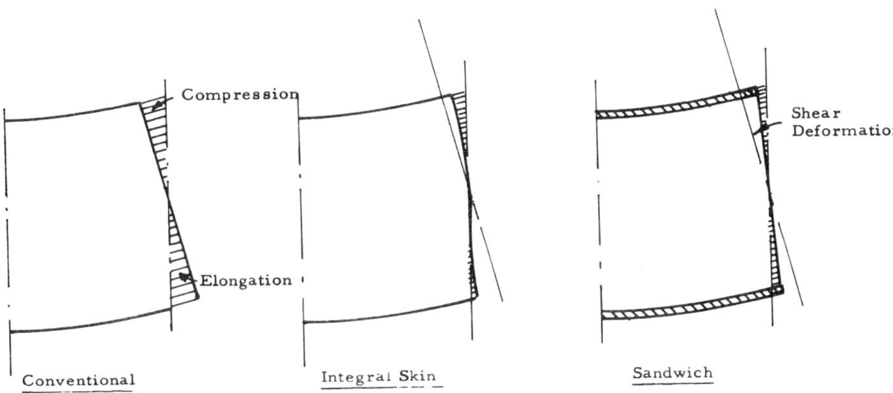

Figure 1. Types of Beams.

Three types of beams are shown in Figure 1. The first is a conventional, homogeneous structure. Stress analysis is, of course, well established. Shear deformation of the center portion is negligible since shear stresses are low compared to tensile and compressive stresses. Compressive and tensile elongation are directly proportional to the distance from the neutral axis and, hence, the familiar $S = Mc/I$.

The second type of beam can be represented by an integral skin rigid urethane foam. While the surfaces are more rigid than the core, the difference is less than in a conventional foam-filled sandwich. The core deforms in shear, but it also contributes in compression and tension to the rigidity of the beam. Structural analysis would be difficult.

The third type of beam is a foam-filled sandwich with thin, flat faces. The core deforms in shear and, since the compressive and tensile moduli are very low compared to the faces, compression and elongation of the core do not contribute appreciably to the rigidity of the beam. Essentially all rigidity comes from the compressive and tensile moduli of the faces and from the shear modulus of the core. The moments of inertia of the faces about their own neutral axis are low and, if the faces were not supported by the core, they would bend easily.

It is these two properties, the low compressive and tensile moduli of the core and the small moments of inertia of the faces that simplify deflection equations to a form starting from:

$$d^2y/dx^2 = M/EI - \frac{dV/dx}{b\,D\,G}$$

Where: d^2y/dx^2 = The second differential of deflection relative to span:

M = Moment, in-lb.

E = Elastic modulus of the faces - psi.

I = Moment of inertia, in^4

dV/dx = Differential of shear force relative to span.

b = Width, in.

D = Panel Thickness, in.

G = Shear modulus of the core, psi.

Design equations are presented in references [1, 2, 3, 4, and 5]. For a uniformly-loaded beam on a simply-supported single span the deflection equation is:

$$y = \frac{5w\,L^4}{384\,E\,I} + \frac{w\,L^2}{8\,b\,D\,G}$$

Where: y = Deflection at the center of the span, in.

w = Load, lb/in of span.

L = Span, in.

As discussed later, urethane foam will creep under sustained load. Creep in effect, decreases the shear modulus in the above equation.

The ultimate strength of a foam-filled beam may be limited by the shear strength of the core. For a beam with thin, flat faces shear stress can be calculated from:

$$Ss = V/b\,D$$

Where: Ss = Shear stress, psi

V = Shear force, lb.

For a uniformly-loaded beam on a simply supported single span the maximum shear force (at the supports) is:

$$Ss = w\,L/^2\,b\,D$$

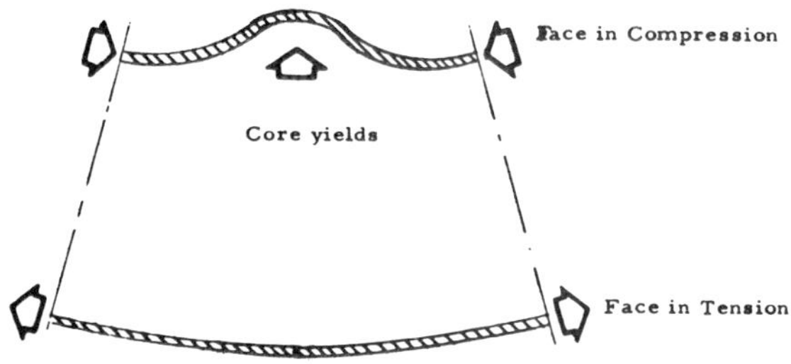

Figure 2. Buckling.

It is more likely that the ultimate strength of a foam-filled beam with thin, flat faces will be limited by the buckling stress of the face which is in compression. For a panel with thin, flat faces the stress in a face can be calculated from:

$$S = M/b\,t\,D$$

Where: S = Compressive or tensile stress in a face, psi

t = Thickness of the face, in.

For a uniformly-loaded beam on a simply-supported single span:

$$S = w\,L^2/8\,b\,t\,D$$

The stress at which buckling occurs in the face is greatly influenced by the rigidity of the core. Short of buckling, the core holds the face straight and allows it to support considerable compression. This is a major reason foam-filled sandwich structures are so strong for their light weight. However, as shown in Figure 2, above a critical load the core yields inward or outward, the face buckles, and the beam fails. Hoff and Mautner, reference 6, developed the following simple expression for calculating maximum buckling stress:

$$Sb = 0.5\,(Ef\,Ec\,G)^{1/3}$$

Where: Sb = Buckling stress in the face, psi

Ef = Elastic modulus of the face, psi

Ec = Elastic modulus of the core in compression or tension, psi (Compressive modulus is lower; use it)

G = Shear modulus of the core, psi.

From the earlier equation for stress in a thin, flat face on a simply-supported single span:

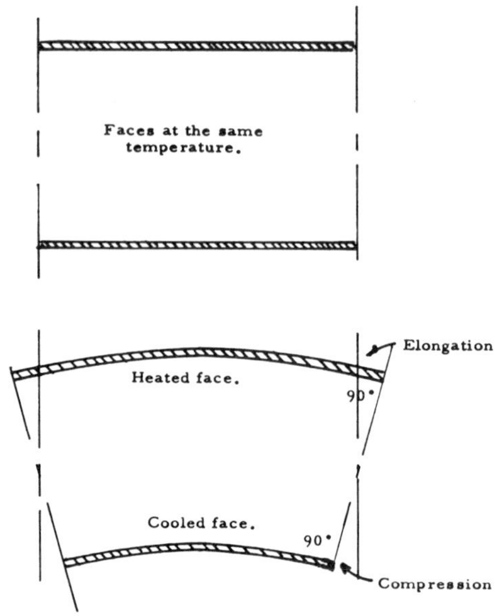

Figure 3. Thermal Wrap.

$$w_b = (4\ b\ t\ D)\ (Ef\ Ec\ G)^{1/3} / L^2$$

Where: w_b = Uniform load required to cause buckling, lb/in of span.

Foam-filled sandwich structures can be deflected by temperature differentials, as well as loads such as imposed by wind and snow. The core is an excellent insulator and the temperatures of the two faces can differ a great deal, even though the panel is thin. As shown in Figure 3, thermal expansion and contraction causes the relative length of the faces to change. In order to accommodate this change in length without causing shear deformation of the core, the panel bows. This bowing is called Thermal Warp. For a panel with thin, flat faces it can be calculated from:

$$y = (k_1 \Delta t_1 - k_2 \Delta t_2)\ L^2 / 8\ D$$

Where: k_1, k_2 = Coefficients of thermal expansion of the respective faces, in/in - °F

$\Delta t_1, \Delta t_2$ = Temperature change of the respective face, °F.

Two additional factors in the design of foam-filled sandwiches are anchoring and joining. They are critical in the development of any structural system and often require more design time than any other aspect. However, anchoring and joining are not well adapted to structural analysis, as are load/deflection and thermal warp. It is only safe to say that the compressive and tensile strenghts of the core are involved. Anchoring and joining devices should tie into the faces, not the core, and loads must be distributed.

FOAM PROPERTIES

The following foam properties have been mentioned as bearing on the structural design of foam-filled sandwiches:

1. Compressive and tensile strength - factors in anchoring and joining.
2. Compressive and tensile moduli - buckling.
3. Shear strength - strength of beams.
4. Shear modulus - deflection of beams.
5. Shear creep - deflection of beams under sustained load.

Ample information is available on foam properties [3, 4, 7, 8, 9]. The difficulty is that most of this information is based on free-rise foams and does not necessarily represent the properties of foam inside a sandwich structure.

Expansion of a rigid urethane foam is complex. Foam is introduced into the cavity between the faces as a liquid or as a froth. As it expands it drags along the faces, gels but continues to expand, and finally hardens into a rigid mass. The cells near the faces become elongated parallel to the faces (parallel to rise). Cells at the center of the core are elongated perpendicular to the faces (perpendicular to rise). Drag along the faces may cause planes of weakness inside the core parallel to the faces. Expansion that takes place after gellation weakens the foam in the direction of rise.

Some of these effects are illustrated in Figure 4, which shows the density and compressive strength profiles for a high rise panel. Overall density and compressive strength perpendicular to rise are relatively constant. Compressive strength parallel to rise decreases steadily from the bottom to the top of the panel, where it is similar to perpendicular strength.

Free-rise data on a similar foam system are also shown in Figure 4. The perpendicular compressive strengths are about one-half the parallel values and are similar to perpendicular strengths in the panel.

Comparative data on a free-rise foam and foam from a horizontally poured two-inch-thick panel are shown in Table 1. The compressive modulus of the panel core (perpendicular to the faces) is intermediate between the perpendicular and parallel free-rise values. Panel shear strengths are only about one-third free-rise. Perpendicular and parallel free-rise shear moduli are similar and also similar to the shear modulus of the panel core.

Tensile moduli and strengths of a free-rise and vertically-risen panel core are shown in Table 2. Parallel free-rise strength is about twice the perpendicular value, while parallel modulus is six times perpendicular. Panel values are appreciably less than even perpendicular free-rise values. Core modulus is two-thirds perpendicular and core strength a fourth of free-rise perpendicular.

One word of caution in comparing the data in Tables 1 and 2. The panel in

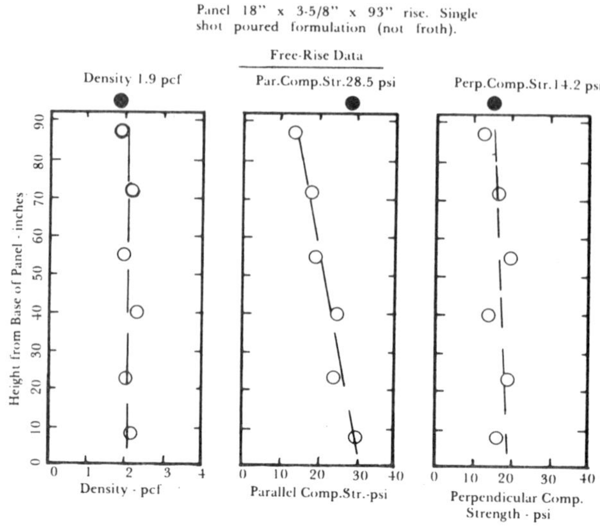

Figure 4. High Rise Panel Core vs. Free-Rise Foam Properties.

Table 1. Free-Rise Compressive and Shear Properties vs. Panel Core [9]

Property[1]	Free-Rise	Panel[2]
Density - pcf	1.95	1.96
Compressive Modulus - psi		
Parallel to Rise	452	365
Perpendicular	174	365
Shear Strength - psi		
Parallel	29.0	
Perpendicular	31.4	11.6
Shear Modulus - psi		
Parallel	205	
Perpendicular	234	201

[1]Panel size 8' x 18" x 2". Horizontally-poured foam with 0.40" aluminum skins.
[2]Panel compressive data obtained with load normal to faces. Shear data taken on samples cut from panel.

Table 1 was carefully prepared by horizontally pouring in such a manner as to minimize flow. The panel in Table 2 was prepared in a more conventional manner with a vertically-risen froth. Table 1 panel data should more nearly approach free-rise values than should the data from the panel in Table 2.

The conclusion of the preceding comparison of free-rise and panel-foam properties is that the best way to get design data on foams is to make a structure and test it. However, free-rise data are usually all that are available. In this case, the following guide lines are suggested:

Density

For similar formulations, packing will cause panel core densities to be higher than the free-rise density of a similar formulation. The degree of packing can only be estimated from experience. The change in properties with density can be estimated from the information shown in Table 3.

Table 2. Free-Rise Tensile Properties vs. Panel Core[1]

Property	Free-Rise	Panel[2]
Density - psi	2.00	1.95
Tensile Strength - psi		
Parallel	55.0	
Perpendicular	29.0	7.6
Tensile Modulus - psi		
Parallel	1,888	
Perpendicular	308	208

[1] Panel size 2′ x 2′ x 3-5/8″. Vertical rise froth. Plywood skins.

Table 3. Corrections For Density [7]

$$\text{Property}_A = \text{Property}_B \left(\frac{\text{Density}_A}{\text{Density}_B}\right)^N$$

Property		Density Exponent N
Compressive Strength		1.4
	Modulus	1.4
Tensile Strength		1.1
	Modulus	1.2
Shear Strength		1.1
	Modulus	1.1

Compressive Modulus and Strength

Compressive loads on a sandwich panel are almost always applied perpendicular to the faces (perpendicular to foam rise). While the cells inside the panel core are oriented parallel to rise near the face and perpendicular to the rise and the center, it is safer and most accurate to use perpendicular free-rise compression data in panel design.

Tensile Modulus and Strength

On the basis of the data shown in Table 2, two-thirds of the perpendicular free-rise modulus could be used in panel design, but only a fourth of the strength. Strength is reduced by weakness inside the core induced by drag of the expanding foam against the faces. Note that any safety factor would further reduce the design strength.

Shear Modulus and Strength

Shear stress is equal perpendicular and parallel to rise. This accounts for the similarity of perpendicular and parallel free-rise and panel shear moduli. Shear strength, however, is reduced by drag of the expanding foam against the faces. On the basis of the information shown in Table 1, one-third of the free-rise shear strength, less a safety factor, could be used in panel design.

Adhesion to Skins

Adhesion between the foam and the faces must be greater than the tensile and shear strengths of the foam. For common types of faces, such as aluminum and steel, this means priming.

CREEP AND STRESS RELAXATION

Foam-filled panels behave well under short-term loads, such as applied by wind to the walls and roof a building, provided the shear stress is limited to a fraction of the foam's ultimate shear strength. Deflection is in near proportion to load, there is little hysteresis loss upon loading, and there is little loss of rigidity upon repeated flexing. Hysteresis loss and loss of rigidity upon repeated flexing increase as stress becomes larger. In one set of experiments there was a ten percent loss of rigidity when a beam was flexed thirty-one times to one-half its ultimate shear strength. A similar loss was noted after ninety-three cycles to only a fourth of the maximum stress. The beams recovered most of their rigidity when allowed to stand without load.

Under sustained load, such as applied by snow to a roof, the core will creep. Creep is initially rapid, then becomes slower with time. When the load is removed, there is immediate recovery (similar to immediate deflection when the load was first applied), and then continued recovery at a continuously-decreasing rate with some permanent set. It is probable that creep of a rigid urethane foam continues as long as the load remains. This limits the use of foam-filled sandwiches in high dead-load situations, such as floors.

Creep is handled in deflection equations by using a reduced shear modulus. Creep rates vary with the chemistry of the foam system. The following equations were determined from the deflection of experimental beams, using two different foam systems, and represent the loss in effective modulus between one day and at least one month.

Slow creep system: $G_t = 0.77\, G_0/t^{0.16}$

Rapid creep system: $G_t = 0.71\, G_0/t^{0.25}$

Where: G_t = Effective shear modulus at time, t, psi.

G_0 = Initial shear modulus, psi.

t = Elapsed time, days.

According to the above equations, shear deformation of the slow system would double in fifteen days compared to four days for the rapid system. One hundred eighty days would be required to triple the deformation of the slow system and 21 days the rapid system.

There is little data in the literature on the creep of rigid urethane foam systems. Some information is given in [10 and 11].

REFERENCES

1. E. W. Kuenzi, "Sandwich Panel Design Criteria", Publication 798, National Academy of Sciences - National Research Council, 1960, pp 9-18.
2. B.H. Fisher, "The Design and Structural Applications of Urethane-Cored Sandwich Panels", Conference on Plastics in Building Structures, The Plastics Inst., June 1965, paper 34.
3. J. M. Buist, H. Gudgeon, "Advances in Polyurethane Technology", Wiley, 1968.
4. D. J. Doherty, G. W. Ball, M. G. Walker, "The Physical Properties of Rigid Urethane Foam Sandwich Panels", Conference on Plastics in Building Structures, ***The Plastics Inst.,*** June 1965, paper 33.
5. J. A. Hartsock, "Design of Foam-Filled Structures", Technomic, 1969.
6. N. J. Hoff, S. E. Mautner, "Bending and Buckling in Sandwich Beams", *Journal of Aeronautical Sciences,* Vol. 15, No. 12, 1948.
7. R. A. Stengard, "Properties of Rigid Urethane Foams", Bulletin of the E. I. DuPont Co., June 21, 1963.
8. B. Hughes, R. L. Wajda, "Plastics Sandwich Panels with various Foamed Core Materials and Their Behavior under Load", Conference on Plastics in Building Structures, *The Plastics Inst.,* June 1965, paper 29.
9. J. A. Hartsock, "Shear and Buckling in Foam-Filled Panels", *Journal of Cellular Plastics,* Vol. 2, No. 6, Nov. 1966, pp 332-334.
10. W. N. Findley, C. A. Stanley, "Nonlinear Combined Stress Creep Experiments on Rigid Polyurethane Foam with Application to Multiple Integral and Modified Superposition Theory", *Journal of Materials,* Vol. 3, No. 4, Dec. 1968, pp 916-149.
11. J. A. Hartsock, "Creep of Rigid Urethane Foam in Shear", *Journal of Cellular Plastics,* Vol. 3, No. 2, Feb. 1967, pp 81-90.

JOHN A. HARTSOCK, *Stran-Steel Corp., Houston, Texas*

CHAPTER 13

URETHANE SEAMLESS FLOORS

INTRODUCTION

The growth of seamless floors has had an exciting and profound effect on both the urethane and flooring industries. To the urethane industry, it has been an important factor in the recognition of the superb properties of urethane films and the rapid growth of moisture curing urethane resins. It has added a new dimension to the flooring industry as a unique flooring concept which can produce attractive and imaginative decorating effects at the job site, by embedding a variety of different colored particles into a liquid polyurethane resin, just poured from a can.

The features of seamless floors are well established. These floors are easily installed, durable, lightweight, flexible, slip and dent resistant, scratch and scuff resistant, stain and dirt resistant, fungus resistant, heel mark resistant, and do not need to be waxed.

While the past decade has established the merits of the seamless flooring concept, the next generation of "Floors From Cans" will require new concepts, new techniques and new resins. To insure its future growth in the flooring market, improvements in color retention, reduction in installation time, reduction in solvent odor and better adaptability to old floor coverings are desirable and necessary.

SEAMLESS FLOORING APPLICATION

Figure 1 illustrates the current method of applying seamless floors. After carefully preparing the floor surface, a basecoat is applied by roller, spray or trowel. This coat is usually pigmented to camouflage voids in the chipping operation.

The chips are broadcast into the wet basecoat and then the layup is allowed to cure. When firm, the excess chips are removed, the layup is sanded to remove sharp edges and smooth the surface, and then vacuumed to remove the sanding dust.

Depending on the amount and type of traffic, two or more urethane glaze coats are applied as the wear surface.

The chips are broadcast either by hand, as shown, or blown on to the wet surface. Blowing distributes the chips more uniformly, particularly on vertical surfaces, and hence, quickly broadens the application range to the hard-to-chip wall and ceiling surfaces.

DECORATIVE EFFECTS

There are a number of colored materials which the imaginative mind can use to provide decorative effects for seamless floors. Commercially, the most

a.

b.

c.

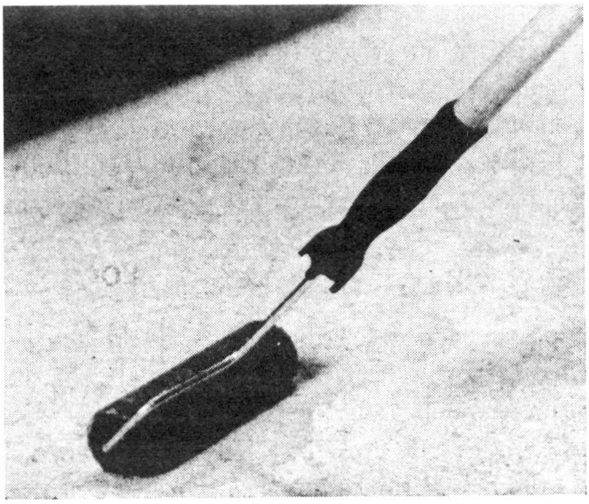

d.

Figure 1. Seamless Floor Application: a) Application of Urethane Seamless Floor Base Coat; b) Sprinkling of Multi-Colored Flakes into Wet Base Coat; c) Sanding; d) Application of Two or More Urethane Glaze Coats.

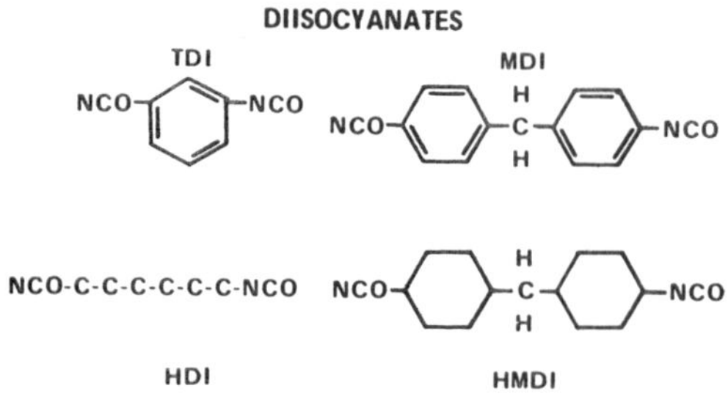

Figure 2.

common material in use is the "vinyl" flake, which is composed of colored and extender pigments bound together, generally, with a polyvinyl acetate emulsion. The flake must be flexible enough to lie flat when applied, friable enough to break up easily and uniformly when tumbling to size, yet tough enough not to powder or crumble.

Colored marble chips and roofing granules are being used more regularly in thinset seamless flooring applications. Although slightly higher in cost, these materials are more durable than the vinyl flake and this is particularly desirable should the urethane be accidentally worn away.

Metallic flakes, available in a variety of colors, shapes and sizes are primarily used for highlighting effects.

Any material, considered for its possible decorative effects in these floors, should not dissolve, swell or curl in the resin, and should not cause any adverse side effects with the resin, such as discoloration, bleeding, and retarded cured.

URETHANE TECHNOLOGY

Polyurethanes are the reaction products of diisocyanates and hydroxyl groups. Figure 2 illustrates the more familiar, commercially available diisocyanates. The monomers in the top half of the figure are aromatic (contain the benzene ring) while those in the bottom half are aliphatic and, as we shall see, the presence of the benzene ring makes a profound difference in some of resins film properties. Of the four diisocyanates shown, most seamless flooring urethane resins are based on either TDI or HMDI. Although pure MDI has been of little interest in seamless applications, Crude MDI, a dark, low viscosity liquid with a functionality of greather than 2, is used as a reactive isocyanate source for high solids pigmented basecoat formulations.

Polyols, the generic name for the hydroxyl bearing compound, are illustrated, according to type, in Figure 3. Polyethers are low in cost and contribute to abrasion resistance. Polyesters are higher in cost and contribute to chemical and solvent resistance. Glycols are used with polyethers and polyesters to impart flexibility. Castor oil and its derivatives are also used with polyesters and polyethers to provide weather and water resistance.

The physical properties of urethane resins are a function of film hardness.

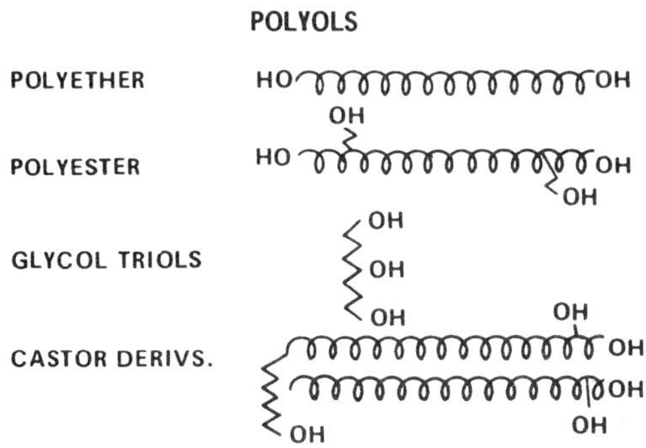

Figure 3.

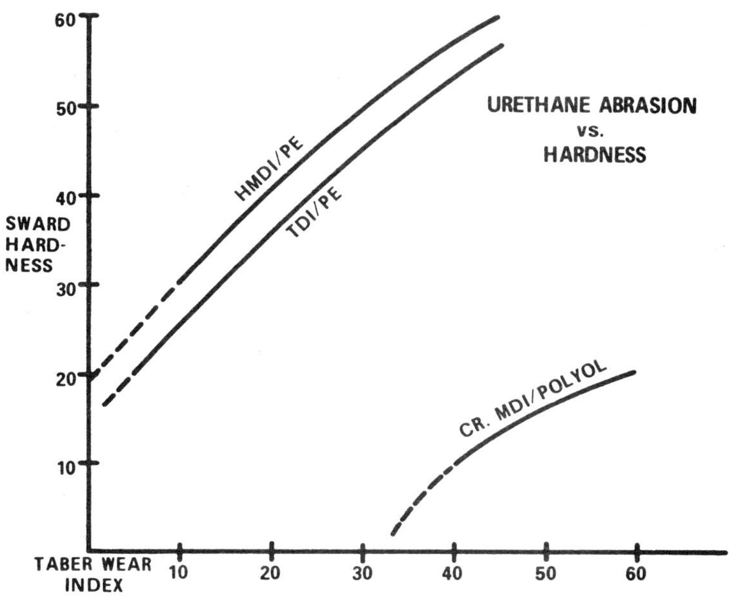

Figure 4. Urethane Abrasion vs. Sward Hardness.

For example, as the hardness increases, tensile strength increases and elongation decreases.

Figure 4 illustrates the effect of the hardness of TDI/PE and HMDI/PE resins on abrasion resistance. In each instance, as hardness increases abrasion resistance decreases. While the wear indexes of HMDI and TDI systems are quite good for any particular hardness, the Crude MDI/polyol systems have a high index of wear, a sacrifice to achieve a high solids liquid system.

The service life of seamless floors has yet to be determined since most of the early floors installed have not failed and are still performing well. The wear resistance of flooring materials can be compared by their resistance to abrasion.

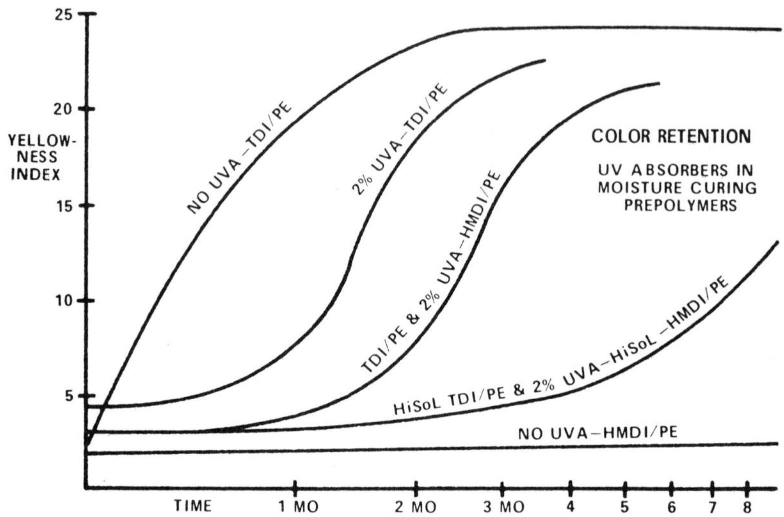

Figure 5. Color Retention Studies: UV Absorbers in Moisture Curing Prepolymers.

Table 1 illustrates a study of the wear index of resilient floor covering materials as compared to aromatic and aliphatic seamless floors.

The aliphatic urethane is an HMDI moisture curing type with a hardness of 50. Except for pure vinyl its wear characteristics were greater than the other floor types. The wear index of the aromatic, a TDI moisture curing resin with a Sward hardness of 22 indicates its service life should be longer than all of the floor types.

COLOR RETENTION

Aromatic urethanes discolor when exposed to an ultra-violet light source, an important consideration when placing floors near picture windows, window walls and glassed doorways. The discoloration is directly related to the presence of the benzene ring in the aromatic urethane. To illustrate, when the unsaturation in the aromatic rings in MDI is removed through hydrogenation, the resulting HMDI is cycloaliphatic and resists discoloration, loss of gloss and degradation of other film properties experienced by MDI when exposed to sunlight.

Ultra-violet light absorbers will delay the discoloration of aromatic urethanes. Figure 5 is a plot of yellowness index vs time of urethane seamless floor discoloration studies run in Florida at 45° South. Three glaze coats on seamless layups were used in this study. The first two curves represent the conventional TDI moisture curing resins and illustrate that 2% UV absorber in each coat will delay the discoloration between 1 and 2 months. Higher levels of absorber have shown even greater effectiveness.

The third and fourth curves illustrate a more efficient and interesting method of delaying the discoloration of aromatic glaze coats. Both curves show the color retentive life of these systems is extended further than the aromatic urethane protected with the absorber in each coat. The fourth curve illustrates that longest protection is achieved if the aliphatic/UV absorber glaze coat is thicker or contains a higher level of absorber. Similar exposures in Buffalo, 45° South and under glass have shown longer effectiveness. After 1 year of exposure, have shown only a slight discoloration while the TDI/UV absorber system has discolored to a deep amber. Apparently when the absorber is present only in the aromatic urethane, the film begins to discolor as the absorber begins to lose its efficiency. When a non-yellowing urethane is used as the carrier for the absorber, the aromatic resin beneath will not discolor until most of the absorber in the non-yellowing topcoat has lost its efficiency. This approach should be of interest where good color retention and minimum cost are requirements.

The final curve is the aliphatic HMDI and it has changed little in color over time.

Table 2 is a list of absorbers studied and found useful with considerations as noted.

Table 1. Taber Abrasion Resistance.

Floor Type	Wear Index*
Aromatic urethane seamless	89
Aliphatic urethane seamless	119
Vinyl Asbestos	147
Vinyl	101
Asphalt	124

*Weight loss method - H-22 wheels.

Table 2.

Cyasorb UV-24(1)	Compatible; efficient; adds considerable color to vehicle
Carstab 700(2)	Intermediate in all three aspects
Tinuvin 327(3)	Compatible in resin solution but tends to "bloom" on surface after 1 to 2 months cure; low color contribution; efficient even when it "blooms"
Tinuvin 328(3)	Similar to Tinuvin 327 in all aspects except broader compatibility range has eliminated "bloom"
Cyasorb 1988(2)	Compatible; very low color contribution; moderate efficiency

(1) Registered tradename of American Cyanamid Co.
(2) Registered tradename of Cincinnati-Milcon
(3) Registered tradename of Ciba-Geigy

INSTALLATION TIME

Since minimum application time under a wide variety of conditions is essential, the dry and recoat time of moisture curing urethanes can be accelerated by amine and metallic catalysts.

Table 3 illustrates suggested levels and properties of catalysts which have been found to be particularly useful. Dabco, a registered tradename for triethylenediamine, has excellent pot life, promotes quick cure, but will accelerate the yellowing of aromatic urethanes. It has no effect on the color retention properties of the aliphatic resins.

Quadrol is a registered tradename for NNN'N' tetrakis-2-hydroxypropyl ethylene diamine. It reduces the working life of the urethane to less than 1 hour. The hydroxyls present, will react with the residual isocyanate groups which results in the formation of urethane linkages, rather than the more color prone urea linkages from the moisture reaction. Thus, the yellowing of the cured films is retarded. Organo-tin and organo-zinc catalysts both maintain work life of greater than one day. Tin catalysts are effective in both aromatic and aliphatic urethane but zinc catalysts have been effective only in the aliphatic urethanes. Tin catalysts will reduce adhesion if used at too high a level.

The aromatic and aliphatic urethane resins currently being used in seamless floor application are the moisture curing types supplied at 40 - 42% solids. Another way to reduce installation time and/or to improve appearance is to apply the urethane resin at higher solids content. Given thicknesses can be achieved in a lesser number of coats or greater build and uniformity with a given number of coats. Urethanes applied at 60 - 65% solids result in 33%

Table 3. Catalysts for Moisture-Cure Urethanes

	(Level on Resin Solids)	Work Life	Dry Time	Remarks
Dabco[1]	0.1 - 0.5%	> 1 day	3/4-1½ hrs.	Accelerates yellowing slightly*
Quadrol[2]	2 - 8%	½ - 1 hr.	3/4-1½ hrs.	Retards yellowing slightly*
Organo-tin	0.1 - 2%	> 1 day	1 ½-3 hrs.	Reduces adhesion at high levels
Organo-zinc	0.1 - 1%	> 1 day	1 - 2 hrs.	For aliphatic urethane only

*Aromatic urethanes only

(1) Trademark of Houdry Chemical Co.
(2) Trademark of Wyandotte Chemical Co.

greater film build per coat. Two glaze coats of these resins would be equivalent, in build, to three coats of the 40% solids resins. Thus, their greater spreading rate results in lower application costs and solvent concentrations.

Table 4 outlines the high solids urethane resins which are particularly useful in high build seamless systems. New aliphatic and aromatic moisture curing resins, thinned to 60% solids with a low odor, high flash mineral spirits type solvent are now available to supplement and complement the standard 60% solids aromatic urethanes thinned with xylol. Because of their low odor they should be of particular interest where the odor of xylol is objectionable. These high solids resins cure too slowly however, to be practical on seamless applications. Thus supplementary catalysts are necessary to achieve satisfactory cure times.

Aliphatic and aromatic adducts, cured with low viscosity polyols, can be applied at the highest solids possible for maximum build in a minimum number of coats. The polyol can be pigmented for use as the base/chip coat. They are particularly useful where high performance coatings are required. Supplementary catalysts can also be used to achieve complete installation in minimum time.

Crude MDI and its derivatives, because of their very dark color, have been used only in base coat applications. Their inherent low viscosities, however, permits the formulation of pigmented basecoats with excellent application viscosities at essentially 100% solids.

USE OVER OLD FLOOR COVERINGS

The ability to apply seamless floors directly over old resilient floor coverings is an important factor in its future growth. In remodeling installations, plywood or hardboard sub-flooring had to be installed or the old tile or sheet goods was removed. Since this added to labor and cost, the development of an easily applied sealer or barrier coat was desirable. The main problems involved are adhesion to the non-porous flooring and bleeding

Table 4. Urethane Resins For Seamless Flooring

Prepolymers	Cure	% Solids	Primary Solvent Aliphatic	Primary Solvent Aromatic
Aromatic/PE	moisture	60	X	X
Aliphatic/PE	moisture	60	X	
Aliphatic Adduct	polyol	60 - 70		X
Aromatic Adduct	polyol	60 - 70		X
Crude MDI derivatives	polyol	90 - 100		
Aromatic Adduct	polyol	60 - 70	X	X
Crude MDI derivatives	polyol	90 - 100		

of colors, mastics and asphalt.

The following is a summary of results obtained with the more successful barrier coats: 100% solids epoxies have good bleed seal to mastics, asphalt and bleeding colors, fair to good adhesion to most surfaces, but require long cures before succeeding coats can be applied.

Polyvinyl acetate and acrylic lactices have been developed which are not very permeable to the solvent in the glaze coat resins and hence, have good bleed seal. Their adhesion to most surfaces is fair but they have relatively poor water resistance.

At 100% solids urethanes have good bleed seal, fair to good adhesion to all surfaces and cure within a few hours to a flexible, water resistant coating.

Epoxy emulsions and latex/asbestos sheeting are new innovations which are still being evaluated for efficiency. To be effective, the barrier sheet must be glued to the old floor hence, another step is involved. The epoxy emulsions have shown some promise, but the available field data is still too sketchy for a proper judgement on its adhesion, bleed seal characteristics and water sensitivity.

THE MARKET

The average annual sales of resilient flooring is approximately 4 billion sq. ft. Vinyl sheet and vinyl asbestos tile are about equal in sales and enjoy about 65% of the total market sales. Asphalt tile has approximately 12% sales, while linoleum and rubber enjoy 10 and 4% respectively. In 1968, about 320 million sq. ft. of urethane seamless was installed or about 8% of the market sales. Installations dropped to approximately 200 million sq. ft. or 5% of the total market sales in 1969. Much of this loss can be attributed to overselling, and poorly installed or underserviced floors. Reputable applicators are becoming established with the result that shoddy applicators are being pushed out of the market.

The growth of seamless floors has primarily occurred through commercial and institutional applications. If we examine the breakdown of resilient flooring sales as shown in Table 5 it becomes quite obvious that this growth has been derived from categories that represent approximately 45% of total

Table 5. Residential Flooring Sales By Market.

		% of Market
Residential		55
Over Counter	20%	
Remodeling	40%	
New Homes	40%	
Public		45
Commercial	46%	
Institutional	43%	
Industrial	11%	

market sales. The residential market which represents approximately 55% of the market is a large untapped potential for increasing seamless floor sales.

A major consideration when purchasing any new product is its cost compared to the value received. Table 6 lists the average installation costs of urethane seamless floors as compared to other resilient flooring materials. Seamless floors are priced in the range of premium sheet goods and thus, would be considered reasonable for a premium product.

One further comparison, which illustrates another "value" of seamless floors, is shown in Table 7. The average maintenance cost of urethane seamless is significantly lower than other flooring materials and is considerably lower than the premium type vinyl and linoleum flooring. Hence, savings realized from lower maintenance costs over the life of the floor should more than offset the installation costs.

Table 6. Average Installation Costs.

	Per Square Foot
Asphalt Tile	$0.35
Vinyl Asbestos Tile	0.45
Vinyl Tile	1.00
Linoleum Sheet	1.25
Terrazzo	2.50
Quarry Tile	4.00
Polyurethane Seamless	1.00

Table 7. Average Maintenance Costs.

	Per Square Foot/Annum
Asphalt Tile	$0.70
Vinyl Asbestos Tile	0.60
Vinyl Tile	0.55
Linoleum Sheet	0.55
Terrazzo	0.40
Quarry Tile	0.40
Polyurethane Seamless	0.18

CARL L. SCHWENK, *Spencer Kellogg, Division of Textron, Inc.*